Riding for the Brand

Riding for the Brand
150 Years of Cowden Ranching

Being an Account of the
Adventures and Growth
in Texas and New Mexico of the
Cowden Land & Cattle Company

Michael Pettit

CC JAL ↳

UNIVERSITY OF OKLAHOMA PRESS : NORMAN

ALSO BY MICHAEL PETTIT

American Light: Poems (Athens, Ga., 1984)
Cardinal Points: Poems (Iowa City, Iowa, 1988)

Library of Congress Cataloging-in-Publication Data

Pettit, Michael.
  Riding for the brand : 150 years of Cowden Ranching :being an account
of the adventures and growth in Texas and New Mexico of the Cowden
Land & Cattle Company / Michael Pettit.
    p. cm.
  Includes bibliographical references (p.).
  ISBN 978-0-8061-3718-6 (cloth)
  ISBN 978-0-8061-4044-5 (paper)
    1. Cowden family. 2. Ranchers—Texas—Biography. 3. Ranchers—
New Mexico—Biography. 4. Ranch life—Texas—History. 5. Ranch
life—New Mexico—History. 6. Ranching—Texas—History.
7. Ranching—New Mexico—History. 8. Cowden Cattle Company—
History. 9. Texas—Social life and customs. 10. New Mexico—
Social life and customs. I. Title.

F385.P44 2006
636'.01'0922—dc22

                                                          2005043938

The paper in this book meets the guidelines for permanence and durability
of the Committee on Production Guidelines for Book Longevity of the
Council on Library Resources, Inc. ∞

Copyright © 2006 by the University of Oklahoma Press, Norman, Publishing
Division of the University. Manufactured in the U.S.A. Paperback published 2009.

In Memoriam

Annie Mae Cowden, 1900–1989
Guy Cowden, 1900–1990

Guy Tom "Rooster" Cowden, 1926–1994
Thomas Cowden, 1954–1973

A. Edwin Pettit, 1914–1998
Suzanne Pettit, 1948–1998

For Emily and Guy

Children were desired and cherished and given all that their families could give, of things and powers and certainties. The blessings of life came from the parents, the ancestors, to whom gratitude and veneration were due. On the hot afternoons of the summer, when the sky blue had golden shimmers over it from heat, and the cottonwoods were breathless and the river ran depleted in and out of their shade, there rose on the hot silver distance the big afternoon rain clouds, with their white billows and black airy shadows. They had promise and blessing in them—rain and life. The people pointed to them and said to the children with love and thanks, "Your grandfathers are coming."

Paul Horgan, *Great River*

Contents

Illustrations

Photographs

Following page 100

JAL outfit mounted, Pecos River country, Texas, 1895
Cowden outfit mounted, Cowden Ranch, New Mexico, May 2000
JAL cowboys, Pecos River country, Texas, 1895
Cowden Ranch branding
Dragging a calf to be branded
Flanking a calf
Working a calf
Branding a calf
Castrating a calf
JAL cowboys castrating, Pecos River country, Texas, 1895
JAL outfit camped, Pecos River country, Texas, 1895
JAL chuck wagon, Pecos River country, Texas, 1895
Taking a break, Cowden branding pens
Another break, Cowden branding pens

Following page 200

Cowden family, Palo Pinto County, Texas, circa 1878
Front gate, Cowden Ranch, Guadalupe County, New Mexico, May 2000
Cowden family, 2000
Jal Cowboy Sculpture, by Brian Norwood, Jal, New Mexico
Painting of Old Adobe in Monument Draw, New Mexico
JAL tally book
Guy Cowden's tally book
Dollarhide oil field, Andrews County, Texas, May 2000

Drawings

Table

Maps

Acknowledgments

Riding for the Brand exists because of the direct and indirect contributions of the Cowden family, past and present. From the time William Hamby Cowden and his brother George Franklin Cowden headed west, first as soldiers, then with families as pioneering ranchers, the family has shown remarkable enterprise and endurance. Today, Sam and Kathy Cowden carry forward a family legacy into the twenty-first century, making adaptations necessary to keep a way of life alive. Their steadfast cooperation has made this book possible.

Equally important are other family members who generously provided encouragement and vital materials: Eugenia Cowden Pettit, Jean Cowden, Mumzy and Walter Kellogg, and all of their children, *mis primos*. As Guy Cowden said about his family, "Twelve grandkids, and not an idiot in the bunch!" Beyond immediate family, scores of other relations and friends opened their doors, minds, and hearts to this project: Simon Barker-Benfield, Bill and Mary Cowden, Frank and June Cowden, James Coley Cowden, Olive Cowden, Vince DiMarco, Bill Jowell, Abby Madden, Jon and Jackie Means, Brian Norwood, Souli and Jake Shanklin, and James Tom, among many others. Thanks also to the cattlemen and cowboys I worked with for demonstrating so clearly their unique character and skills.

In the course of research, many individuals and institutions were very helpful: Pat McDaniel and Jim Bradshaw at the Haley Memorial Library and History Center; Alfred Bush, Curator of the Philip Ashton Rollins Collection, Firestone Library, Princeton University; George Miles, Curator of the Yale Collection of Western Americana, Beinecke Library, Yale University; the Amherst College Library; the American Antiquarian Society; the Panhandle-Plains Historical Museum; the Southwest Center/Special Collections Library of Texas Tech University, Betty Orbeck of the Petroleum Museum in Midland, Texas; and Nancy McKinley of the Midland County Historical Society. Also I thank Rosemary Wetherold for her exceptional editing of the manuscript.

Above all, I thank my daughter, Emily, and my son, Guy, for providing the inspiration to tell this story. In all ways they carry me forward.

Riding for the Brand

COWDEN RANCHES
IN
TEXAS & NEW MEXICO
1853–2005

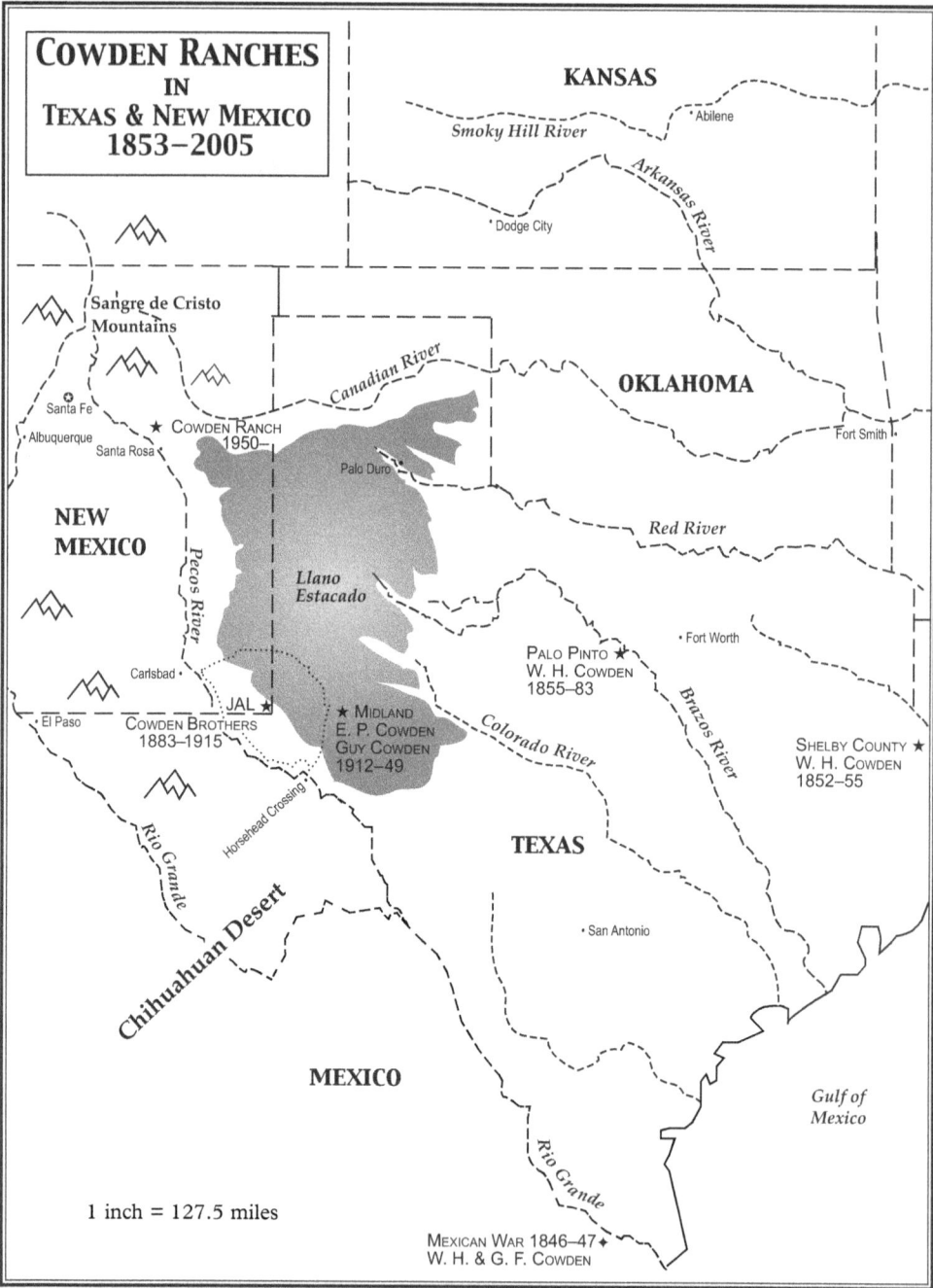

KANSAS

Smoky Hill River

• Abilene

Arkansas River

• Dodge City

Sangre de Cristo Mountains

Canadian River

OKLAHOMA

✿ Santa Fe

Albuquerque

Santa Rosa

★ COWDEN RANCH
1950–

Palo Duro

Fort Smith

NEW
MEXICO

Pecos River

*Llano
Estacado*

Red River

• Fort Worth

PALO PINTO ★
W. H. COWDEN
1855–83

Brazos River

Carlsbad •

• El Paso

JAL ★
COWDEN BROTHERS
1883–1915

★ MIDLAND
E. P. COWDEN
GUY COWDEN
1912–49

Colorado River

SHELBY COUNTY ★
W. H. COWDEN
1852–55

Rio Grande

Horsehead Crossing

TEXAS

Chihuahuan Desert

• San Antonio

1 inch = 127.5 miles

MEXICO

*Gulf of
Mexico*

Rio Grande

MEXICAN WAR 1846–47 ✦
W. H. & G. F. COWDEN

Roundup

Sam Cowden is horseback, on Hollywood—"best horse I ever owned." A neighbor strapped for cash sold Hollywood—best horse *he* ever owned—to Sam because Sam would ride him right, give him a good home. And Hollywood has the best home a horse might imagine—the Cowden Ranch, a fifty-thousand-acre family ranch near Santa Rosa, New Mexico.

Hollywood, a handsome sorrel gelding, is the horse Sam rides in rodeos, in team roping competitions, but he's more than a performer. Hollywood's a working horse, and today Sam is using him to gather and sort cattle in Conchas, a seventy-five-hundred-acre pasture on the northern part of the ranch. Early this morning Sam and a handful of cowboys left Cowden headquarters, trailering their horses to the far reaches of Conchas, about five miles away, where they began their roundup. "Roundup" and "rodeo" both derive from the Spanish verb *rodear,* "to surround" or "go around," and modern rodeo—with its bronc and bull riding, its clowns and announcers, its team and calf roping in big arenas before cheering crowds and television audiences—derives from the humble and solitary work of cowboys on horseback, like the roundup today.

It's beautiful country where Sam ranches—Sangre de Cristo Mountains rising snowcapped to the northwest, Conchas River off to the north, and isolated mesas dotting the southeast, where the Llano Estacado, the fabled Staked Plains, stretch out forever. It's beautiful, spare, open country, where the wind almost always blows, as it does this morning. Sam and the other riders spread out around the pasture, keeping each other in sight as specks on the horizon, and everyone understands his place, in the roundup and in the world. Land and sky and wind keep each cowboy honest.

As morning progresses, the cowboys sweep back and forth around the perimeter of the pasture in a tightening noose, picking up cattle along the way, pushing them toward a water tank where they will rendezvous. The riders watch one another's distant movements and react accordingly, in concert, keeping space between them but allowing no cattle to slip through. Speech is impossible; the only sounds are the wind and windblown bawling of cattle. Eventually, as the circle closes, riders can hear one another whistling or yipping at the cows and pokey calves they drive ahead of them.

The cattle have been through this before—Sam brands his calves every spring, early in May—so they know the drill. They know the tank where the roundup, or "gather," converges, if not its name—Big Red. There was once a windmill at Big Red, but now water is pumped to metal storage and watering tanks where the cattle collect. When the gather is complete, each cowboy arriving with his group of twenty to thirty head, it's time to sort the herd.

Sam is careful not to overgraze his land and keeps only about 150 head of cows—fifty acres of pasture for every cow—with their new calves, in Conchas. Each year, however, some cows do not deliver calves or calve late; those "dry" and "heavy" cows must be separated from the cows with calves. Before doing anything, however, the half-dozen riders around the herd sit patiently on horseback and give the cows time to "mother up," to find their calves in the milling herd, the pairs recognizing one another's scents and voices through the din. Once that is done and the herd settles and grows quiet, Sam and Hollywood begin their work, sorting.

Horses, like riders, have certain gifts and learned skills; Hollywood and Sam are a well-matched pair. Both have been on ranches all their lives, have received good training, and attend to their work diligently and intelligently. "If you don't know what you're doing with a herd of cows," Sam says, "it's easy to cause a wreck"—scatter a herd just gathered, run a heifer through a barbed-wire fence, lose a whole morning's work through inattention or bad judgment. Some cowboys love a wreck; it's in their blood, a hot spot they find thrilling, like a bar fight. Not Sam, whose temperament and methods are as calm as possible. Knowing how and when to apply pressure is critical in working cattle; good cowboys like Sam and good horses like Hollywood have a light touch.

Sam walks Hollywood into the herd, looking for any neighbors' cattle that might have strayed onto the ranch. "Code of the West," Sam says of a practice that began on the open range and continues today, even though the Cowden Ranch is well fenced: when working cattle, you always first check for outside brands. Sam finds no strays he would need to return to their proper range, so he begins looking for dries and heavies to cut out. A cow in good condition, fat since no calf is taxing her, her udder not swollen with milk, catches Sam's eye—a dry. He puts Hollywood on her tail—a slight neck rein or pressure with his knees is enough—and with only a turn or two, Hollywood knows which cow to cut out. Horse and rider almost imperceptibly push the cow to the edge of the herd, where the other cowboys wait and watch. They move their own horses to the side, then forward, to give the cow a path out, but no way back into the herd. Only if a cow balks are quick movements required by horses and riders— a good horse will head off a cow before the rider ever gives a signal—but most often the culls are quietly separated. Once a few cows are cut away from the herd, others are less reluctant to follow suit, and the sort goes smoothly, with the qualities of a formal dance—cowboys, horses, and cattle all have their steps, their turns and counterturns and stands, like a minuet. To discover such surprising elegance, in such beautiful surroundings, is eye-opening. Our notions of cowboys often run toward stampedes and gunplay, toward dramatic rather than subtle behaviors, but movie stunts don't play often on a well-run ranch. Home on the range in the year 2000 is not what it used to be; it's what it was, and much more.

In his eighties, Guy Cowden would joke to his grandson Sam Cowden, "I'm nothing anymore but a rhinestone cowboy." Born in 1900, Guy lived out the transition of the West from open-range ranching to modern corporate agribusiness and hobby ranchettes. Barbed wire and windmills appeared, with homesteaders not far behind; railroads replaced trail drives; the Dust Bowl and oil boom came and went, etching their effects into the landscape and culture; isolated ranches merged into a worldwide web; cowboys onstage supplanted cowboys in the saddle. The twentieth century must have seemed terribly chaotic.

When Guy was born, Cowdens ran one of the renowned free-range ranches in the Southwest—the JAL Ranch, which at its height included some forty thousand head. With no fences to hold them, JAL cattle scattered over seventy-five hundred square miles of territory, restrained only by the Pecos River, which loops around the western and southern reaches of the range. According to W. C. Cochran, who worked for the Cowden Cattle Company in the late 1800s, "Their cattle ran all the way from Midland to Carlsbad in New Mexico and from Horsehead Crossing to the south." Located where the southern Great Plains meet the extreme northern reaches of the Chihuahuan Desert, the country is semiarid grassland, and cattle scattered widely to find adequate grass and water. Winter storms and summer droughts also drove cattle in search of shelter and nourishment, with cowboys trailing after on horseback. The JAL Range intersected that of other outfits like the San Simon, Mallet, and

Hat ranches; "reps" from area ranches would participate in general spring and fall roundups to sort out their various cows and calves. Gathering cattle left to their own devices for half the year, cowboys crisscrossed the mesquite-studded plains, sorting and branding before returning livestock to their proper ranges. It was, however, the end of an era.

What had been open range saw increasing development and regulation, and in 1912, when New Mexico finally became a state, one of the older Cowden brothers bought out the interests of the others—including the youngest, Guy's father, Eugene Pelham Cowden. The Cowdens dissolved the JALs—as the JAL Ranch was often called—and, like the cattle they had long herded, scattered throughout the territory. Leasing or buying their own individual ranches, Cowdens went on with their life work—if smaller in scale, less romantic in spirit. By the time Guy Cowden was of age, with a family and a herd of his own, things were changing rapidly, though change wasn't always evident out on the ranch, riding horseback to check windmills or calving cows. He faced the same wind and sun and dry land, the life was still demanding, the country was still poor. On the surface.

Beneath the harsh landscape of West Texas and eastern New Mexico, however, lies the Permian Basin, containing enormous reservoirs of oil and gas, which grew by 1950 into the leading oil-producing region in the United States. Ranchers like Guy Cowden, who had struggled through the Great Depression and the Dust Bowl on as little as six hundred dollars a year, suddenly saw their fortunes rise. Oilmen poured into Midland, Texas, where Cowdens were a fixture, and drilling rigs sprang up everywhere, towering over the largely treeless landscape. Money from oil leases and royalties was suddenly available for fences and windmills, for horses and cattle, for more and better land. And for rhinestone suits. Working cowboys climbed down from saddles that had worn them out, and pretenders took up residence, only looking the part. "Big hat, no cattle" goes the derisive western phrase, and contemporary cowboy singers are called hat acts.

That was never the case for Guy Cowden, who always had a ring of sweat around his Stetson. As the years passed and his own abilities slipped away, he nevertheless kept alive both the ideal and the actual cowman's life. He bought a new and promising ranch up in New Mexico, stocking it with good Hereford cattle, a legacy for his family and for his life's work. When he died in 1990, the Cowden Ranch was producing top-grade calves every year, and his whole family shared in its success. Coming from all over Texas and from New Mexico, Louisiana, California, and Massachusetts, children and grandchildren gathered back in Midland for his funeral, each of them with stories to tell, everyone aware of what was lost and what remained.

William Hamby Cowden, Guy's grandfather and the first Cowden to settle in Texas, left a cold trail behind. When tracking him and other family from generations back, there's little to go on—no letters survive, few first-person accounts. Like the Indians they helped drive from the frontier, they wrote almost nothing down—theirs was an oral tradition or perhaps a tradition of silence. In the Palo Pinto country of Texas in the 1850s, silence was a virtue. Back then, when Comanche raiders might be anywhere, silence could keep you alive, although it threatens to keep you from coming to life now, 150 years later.

Some accounts and records of W. H. Cowden remain, collected in county histories and courthouse archives, facts or purported facts that flesh out his bones: date and place of birth, military records, marriages and children, gravesite. We can place him in a period and landscape, in a culture, letting history and imagination shape a life for us, probably as adequately as lives we know firsthand. From the silence, we can gather shadows.

Bill Cowden was an accomplished cowman, though in ways no modern rancher practices. Moving with his family first to eastern Texas, then to the frontier west of present-day Fort Worth, he was always looking for something—adventure, opportunity, land, a home. The one they made on the Texas frontier was difficult and perilous, situated in Comanche territory that Texans were trying to "settle" and could not, until 1874, when Indian resistance there was finally crushed. For years white settlers and Indian raiders swapped outrages, stealing land or horses from one another, inhabiting the same landscape but quite different worlds. At times, though, whites and Indians seemed like one silent shadow.

Moonlit nights in Palo Pinto were always cause for vigilance, and often alarm. It was then that parties of Comanches or Kiowas would slip into the country from Indian Territory to the north to steal horses and cattle. Whites were intentionally and systematically exterminating the buffalo on which Plains Indian cultures depended, with the aim of conquering the Indians and opening the territory to settlement. Naturally, Comanches and other tribes responded with raids on settlers; Bill Cowden and his neighbors would regularly find, mornings after a full moon, their horses and cattle gone. "The Indians stole out the two old Cowden men, Uncle Billie and Uncle Frank, they kept them stolen out all the time so they did not have any horses to maverick on," wrote W. C. Cochran. (Unbranded cattle that ran wild, particularly during and after the Civil War, "mavericks" were fair game for anyone with a branding iron.) So the Cowdens "commenced their mavericking days barefooted and afoot," Cochran noted. "They would round up a bunch of old poor cows and calves, run them down and what they could not pen they would catch with their dogs."

Running down wild cattle on foot perhaps appealed to Bill Cowden for its adventure, and certainly for its profit. The same was true of later roundups of "wild and outlawed cattle," Cochran said. "These cattle were gathered in this way. One outfit of men would gather forty or fifty head of gentle cattle and find out where a bunch of wild cattle were located and wait until the moon was shining all night. Then they'd drive these forty or fifty head of gentle cattle to the prairie or some open place where these wild cattle would come out of the mountains and brush to graze, [and] round them up with the gentle cattle. You can drive off any wild bunch of cattle after night if you can get them rounded up with a gentle bunch. Old man W. H. Cowden has been on many a night drive. . . . He always knew more where these wild cattle were located and how to get to them. He would go just for the fun of rounding up the outlawed steers and cows." A silent shadow perhaps, but William Hamby Cowden was a frontier cowman of the first order, like his well-known Palo Pinto neighbors Charles Goodnight and Oliver Loving. All of them left their marks, their legacies.

Tumbleweeds

Branding 2000

May 1, 2000—This morning I leave my house in western Massachusetts and drive an hour to Hartford, Connecticut, where I board a Southwest flight for Chicago, with a connection to Albuquerque. In my bags are boots and spurs, blue jeans and shirts with pearl snap-buttons, bandannas and leather gloves. You can see by my outfit that I am a cowboy. Also packed are my laptop, tape recorder, notebooks, and cameras, since I'm on assignment—on the trail of real, practicing cowboys and cattlemen. Friends in Massachusetts listened with romantic envy to my plans.

From Albuquerque Sunport, I drive two hours to Santa Rosa, New Mexico, then another hour on an unpaved ranch road to my destination— the Cowden Ranch, where I will help my first cousin Sam Cowden work his cattle. My trip covers more than two thousand miles and takes fourteen hours—a long day of travel, but only one day. When the Cowdens originally made their way across the United States to Texas and New Mexico, they had countless fourteen-hour days, and it took them generations to make the trip. In comparison, my journey is a jaunt, a little two-step to the West.

As I began to assemble the facts and fictions for this book—about cattle ranching in the Southwest, about a family that seemed like a permanent fixture of that life and landscape—I was surprised to discover that the first Cowdens in America had settled in North Worcester, Massachusetts, less than an hour from where I now live. Looking through a family history in the stately reading room of the American Antiquarian Society library, deep in the heart of New England, I was not far from my roots, as I had thought, but nearby—Massachusetts was home almost three hundred years ago. But not for long. The Cowdens seem to have been a restless clan, moving first as MacCowdens from Scotland to Northern Ireland in the seventeenth century, then joining other Scotch-Irish immigrants looking for better prospects in the New World. Once here, in 1730, like many human

COWDEN RANCH
SANTA ROSA, NEW MEXICO
GUADALUPE & SAN MIGUEL COUNTIES

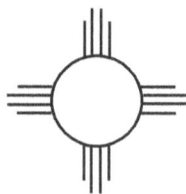

Fences: solid lines   Roads: dashed lines
Acreage: numbers

■ Tank   ✣ Windmill   ● Spring   ❖ Pens

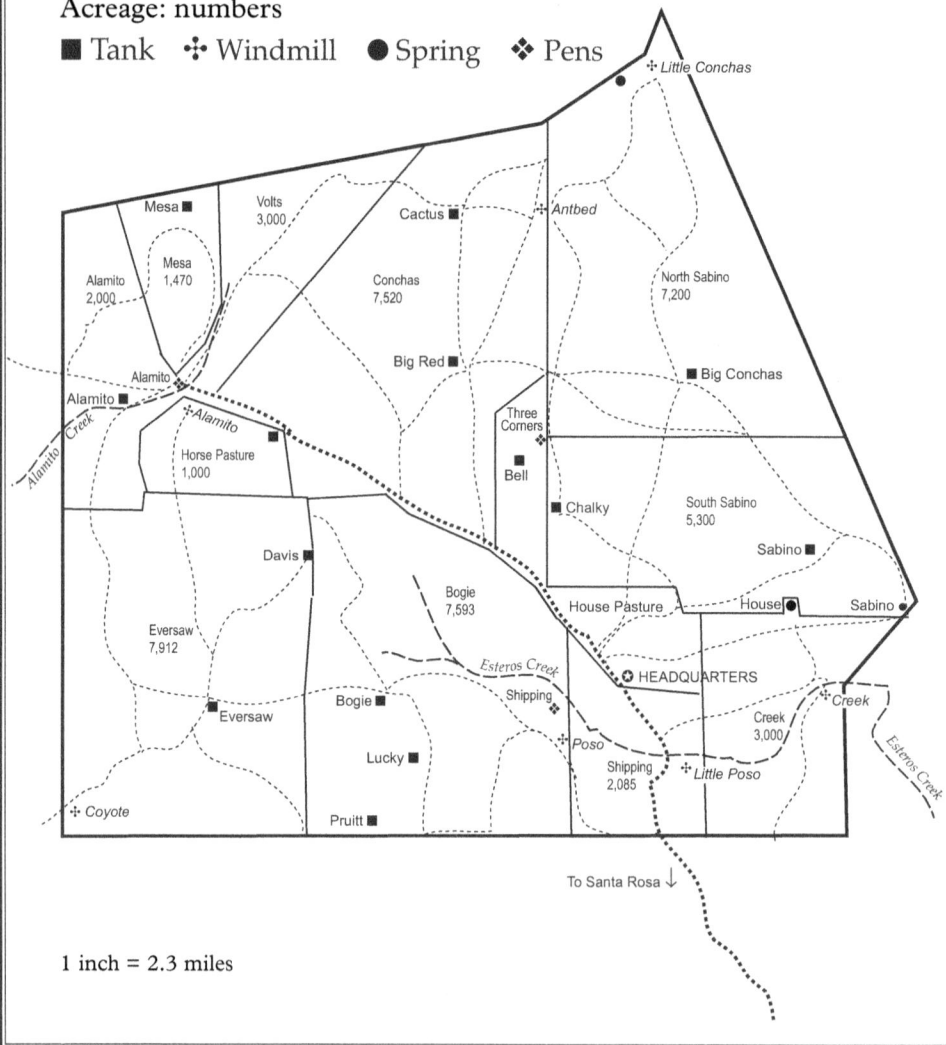

Little Conchas

Mesa ■

Volts
3,000

Cactus ■

Antbed

Mesa
1,470

Alamito
2,000

Conchas
7,520

North Sabino
7,200

Alamito

Big Red ■

Big Conchas ■

Alamito ■

Alamito

Three
Corners

Horse Pasture
1,000

Bell

Chalky ■

South Sabino
5,300

Sabino ■

Davis ■

Bogie
7,593

House Pasture

House ●

Sabino ●

Eversaw
7,912

Esteros Creek

Shipping

HEADQUARTERS

Creek

Bogie ■

Poso

Creek
3,000

Esteros Creek

Eversaw

Lucky ■

Poso

Shipping
2,085

Little Poso

Coyote

Pruitt ■

Alamito Creek

To Santa Rosa ↓

1 inch = 2.3 miles

beings and Americans in particular, they have remained on the move: Massachusetts to Virginia, Carolina, Georgia, and then Alabama. After their long, intermittent migration down the eastern seaboard, in the mid-1800s they headed west in a hurry, scratching some traveling bug.

Except for Sam Cowden, who says agreeably, "I never go anywhere." Although responsibilities keep him close to home, could it be that Sam has found what all those Cowdens were looking for—the right place to make a stand, good land to raise cattle and a family? Could Sam have *arrived*? Could he be satisfied on the ranch as nowhere else? Born in Santa Rosa, raised on the ranch, Sam went off to Eastern New Mexico University, down in Portales, then came back for good. He and his wife, Kathy, have three children, now the fourth generation to call the Cowden Ranch their home. For more than forty years Sam has watched the sun rise over the distant Llano Estacado and set over the Sangre de Cristo Mountains, raising thousands of calves—forty generations!—under the wide, clear skies of New Mexico. To wonder why Sam stays put is to wonder about all that brought him here.

What in William Hamby Cowden's blood drove him to Texas? What sent him off into the Palo Pinto brush to round up wild cattle? That drive endured in his grandson Guy Cowden out in West Texas and endures in Guy's grandson Sam Cowden in New Mexico. Some constellation of landscape and labor and character has kept the Cowdens at home on the range for one hundred and fifty years now, and the stars remain both bright and mysterious. Today Sam carries forward what was set into motion six generations ago; at the same time he steadfastly resists change. What keeps him faithful to that work, that world?

I myself have Cowden blood, I was born into the Southwest, and I've raised cattle, like my grandfather and my cousin. But I walked away from ranch life, although I've kept it alive in my imagination, which may be the only place it can finally survive. I hope not. I hope that ranching remains an actual as well as mythic life, that it has a future as well as a past.

Each summer of my childhood, we left Louisiana, headed west. By car, occasionally by train, we rolled past mossy swampland, sugarcane plantations, rice farms—verdant landscapes we were eager to put behind us. We looked ahead, toward Texas, trying to spy the first cactus, first jackrabbit, first cowboy on horseback. We'd cross the state line, and little would change— we could get Dr Peppers there, but the world appeared much the same: live oaks and piney woods, little towns and cities. It was not yet the West.

If we took the train—my mother, sisters, and I cozy in our Pullman sleeper—we would wake after a night's clattering dreams to find ourselves in another world. If my father was with us and we drove, the transformation took longer; we went to sleep at a motel somewhere in East Texas and woke in the same place. Eventually, past Dallas and Fort Worth, the country opened up: mesquite and cactus replaced green trees, Hereford cattle grazed rangeland, the sky stretched blue to distant horizons. We still had far to go—my grandparents lived in West Texas, and the ranch was even farther, in New Mexico—but once the plains began, we felt we had arrived. We felt at home, almost.

Never did time drag so heavily as the last hours we drove west on Route 66 toward Santa Rosa. Burma Shave signs might distract us, flashing past, and billboards for the Club Cafe promised we would get there sometime, but when? Each rise brought only another long stretch of highway crossing open country, more cactus and mesas and sky. Where was our turn?

To get to the Cowden Ranch, I came to learn over the years, you leave Route 66—now Interstate 40—just before reaching Santa Rosa, located between Tucumcari and Albuquerque. You turn north, passing under the railroad tracks parallel to the highway; once on the dirt road, you'll find rangeland and cattle but few human beings. In the last ten years a handful of ranchettes have sprung up near the highway, like weeds along a ditch, but beyond them you're deep in cattle country. Most of the land once belonged to the Bar Y Ranch, operated by the Driggers family. As children, my sisters and I would ask at each cross fence we encountered, "Are we there yet?" and the answer always seemed to be, "Nope, still the Bar Y."

Now the Driggers family is gone, and Henry Singleton, who rivals Ted Turner as the largest private landholder in New Mexico, owns the Bar Y. It's twenty-six miles from the railroad back to the Cowden Ranch, virtually all of it across Singleton land. The road has changed, improved by cattle guards at the cross fences, where once we had to stop, open a barbed-wire cattle gap, drive through, and close back the gap before driving on. Our eyes were peeled for rattlesnakes, and we had to remember not to shut ourselves on the wrong side of the gate, like silly city kids, but we'd fight for the right to open gates. Anything to get there!

And everyone wanted to be first to see—in the distance, tucked against the mesa bluff—the red roof of the Cowden Ranch. There on the front porch our grandparents waited, watching the horizon just like us, looking for dust from our car. They always bet one another a cold Coke on what time we would arrive, impossible to know since there was no telephone line to the ranch and certainly no cell phones to keep in touch. Like ranch folk for generations, they stared across the open country, looking for family to arrive from a great distance. No wonder the welcome was always so warm.

Even now I can hardly wait.

May 2, 2000—I'm in the bunkhouse, body clock still on eastern time and up early, before the chickens, though there are no chickens. When I was a child, roosters would wake us, but it's quieter here now, a bobwhite quail whistling from the cedars, a lone calf bawling in the distance. I remember it differently.

Used to be, before 1979, when Farmers' Electric finally strung a line out to the ranch, a diesel generator rumbled day and night in the powerhouse out back, an unnatural sound this far from "civilization." The old blue Kohler Electric Plant sits idle in the tin garage, now called the Bullpen, since it's where Sam and Herman Martinez and others repair at the end of the day to drink a beer and bullshit. Inside the Bullpen is a refrigerator stocked with Miller and Coors Light beer and bottles of cattle vaccine. Worn armchairs offer humble, welcome comfort after hours in the saddle. Bleached deer and antelope horns and cattle skulls glow on the dim walls. In back sits the old chuck box off a wagon; a long workbench holds a jumble of tools; in one corner a fifty-year-old Maytag wringer washing machine is surrounded by cast-off parts from untold machinery. Like many places out in the country, materials accumulate over the years for that moment they may be needed. You don't dash to the hardware store from here.

The powerhouse frightened me as a child, since I was warned repeatedly to stay clear of the big, loud generator with its whirling flywheel and belts. Jean Cowden, whose husband, Guy Tom—"Rooster" to everyone who knew him—took over the ranch from my grandfather, remembers when they first got electricity: "We couldn't sleep through the night. We'd wake up and couldn't hear the generator. Or we couldn't hear the generator and would wake up. That made us uneasy. Before, anytime it stopped running, it meant trouble. Took us the longest time to get used to the quiet. Isn't that strange, in the country?"

Cowden headquarters is what in Spanish they call the *hacienda,* a word that derives from the verb *hacer,* "to make" or "to do," and properly refers not to a ranch house alone but to the entire ranching operation, all its diverse labors and lands and structures. Besides the Bullpen, out back is the old cinder-block barn and saddle room, surrounded by corrals. The saddle room with its racks of saddles and blankets and bridles, redolent of leather and horse sweat, opens onto the corrals where horses are fed daily and caught to ride. Between saddle room and the barn are two wooden stalls, once used for milk cows now long gone from the ranch. Inside the barn are salt blocks, tubs of minerals, and other supplies, where hundreds upon hundreds of fifty-pound sacks of cattle feed once rose to the ceiling, subject to loss from rats and mice, laborious to unload and load by hand. Today Sam uses a new set of feed bins, five elevated hoppers that each hold ten tons of "range cubes," delivered directly into the three-quarter-ton feeder trucks used around the ranch. "Pretty slick," says Sam, who can drive under a hopper, pull a lever, and fill his truck in under a minute. No rats, no handling, no problem. Next to the feed bins, an open-sided barn provides cover for horse trailers and a supply of hay bales for the horses.

Also new is a large metal barn that houses various ranch equipment, from the road grader to campers to tractors, and serves as the welding shop, where equipment can be repaired and cattle guards and gates fabricated. Most of the pens and corrals on the ranch are now made of welded pipe or surplus highway guardrails—stout stuff that never requires maintenance, though "something's always tearing up around here," says Sam.

The old chicken house where we used to collect eggs stands empty, beside pens built for my grandfather's hunting dogs. Guy Cowden was an avid hunter, chasing coons and bobcats and mountain lions with his dogs. I can't recall all the different bluetick, redbone, black-and-tan, and other hounds Big Guy had at the ranch. They lay around in dusty wallows in their pens, waiting for the evening he would load them into the truck and head for some watering hole where they might pick up the trail of a raccoon and run bawling through the night. Big, loose, and mild when they weren't hunting, his dogs had a place anywhere Guy did. Always in the kitchen was a scrap bucket filling with treats for Joe, Minnie Pearl, Red, and Tooter; copies of *Full Cry* and *Field and Stream* lay about the house. Nearby are two large pens used for isolating animals: calves Herman is raising, a young colt or two, an injured horse that needs doctoring every day. "Hospital pens," says Sam.

Herman's house—a three-bedroom trailer nicely landscaped with trees to break the wind and dust and sunlight—faces west, toward the barns and corrals. Herman and his wife, Debbie, live with their son Ricardo; their older son has moved away. Herman grew up west of Santa Rosa in a large family, and you don't hear him complain about his current home, or much else for that matter; he seems perfectly at ease and at home, as you would hope after living and working here for twenty-eight years. Over at Alamito is another house, moved there by the Cowdens from a site down on the Pecos River and now used by Juan Estrada or other ranch hands who have come and gone. Scattered around the ranch are ruins of old rock houses built by Hispanic sheepherders in the nineteenth century, evidence of the long history of this hacienda.

The ranch house where Sam, Kathy, and their three children live has changed over the years, as various generations have taken it over. When Guy and Annie Mae Cowden first bought the ranch, no structures existed anywhere except for an old line camp over at Alamito, once used to house cowboys working far from ranch headquarters. Guy and "Mushy"—as Annie Mae was called by her family—built the ranch house, from plans my mother drew, nestled against the mesa and protected from west and north winds. A kitchen–living room and three bedrooms all opened onto a long, deep front porch facing south, where they could see the front pastures and the road to town. Guy and Mushy spent evenings after supper on the porch, watching clouds dissolve, listening for coyotes, waiting to retire for the night. In the mornings the porch, with its cement floor, was both cool and sunlit, a place where they could drink coffee as dawn broke. Attached to the house was a carport, soon converted into living space, and beyond that, another small bedroom and bath—servants' quarters.

Now the whole south wing of headquarters has been incorporated into living space. There's a small office on the west end, a master bedroom, a laundry and mudroom, the same kitchen and living room at the heart of the house, three bedrooms, and the porch, now enclosed. Additional space was necessary because Kathy homeschools the three children. The ranch house is in constant use, the center of the Cowdens' working and private lives, although frequently everyone is outside, where fifty thousand acres of land calls to them.

The wing attached to the ranch house includes a large carport and the bunkhouse, with a big kitchen, two bedrooms, a bath, and a sleeping porch. Over the years, different people have lived there, from the Romero family when I was a child, to Sam and Kathy after they were first married, and now visiting friends or cowboys or hunters. The bunkhouse kitchen is spare but always busy, with an antique Garland stove with six gas burners and a built-in griddle—a big black magnet for anyone hunting a cup of coffee or a meal. When cowboys stay at the ranch during branding or shipping seasons, the bunkhouse becomes the center of activity; Sam comes over from the house to eat with the crew and talk over all manner of subjects. Kathy rarely appears. "The bunkhouse is *no* place for a woman," she says good-naturedly. On her own family's ranch there was apparently greater separation between male and female spheres than is generally common with the Cowdens. Her point is nonetheless well taken; a bunkhouse full of cowboys can get rank, and most women would either dampen or suffer from the atmosphere.

The ranch is reasonably modern, much updated from fifty years ago, when there were stockade corrals built of cedar posts, windmills, no electricity, no telephone. But even today's improvements—from automated feeders to Internet access to cell phones—exist in a very isolated atmosphere. We are still twenty-six rough miles from pavement, thirty from a post office, ten from the nearest neighbor. In *Great River,* Paul Horgan wrote about the haciendas of early Spanish settlers along the Rio Grande: "Each clustered around the life of a family that had chosen to live away from the mountain capital of Santa Fe. Each was a walled sanctuary against the dangers of distance and unconverted Indians. Widely separated from its nearest neighbor, the estate had to be self-sustaining in all things."

Indians are no longer a threat, but the Cowden Ranch nevertheless evokes an existence centuries old. A bird's-eye view looks down on a small compound of houses and barns and pens almost lost in the landscape. I look out from the Cowden ranch house and see open country, high plains unmarked by more than a windmill with a handful of trees. Scattered throughout the plains and mesas are cattle and antelope and coyotes, but no human presence beyond the isolated rock ruins. Thank God we are mostly alone out here. That hasn't changed, and won't for a long time. As the world fills up, and cities spill their millions out into the surrounding countryside, and nothing much remains beyond overwhelming human influence, I hope this land will linger as remote and open. There's not enough water to sustain dense populations of *anything,* although that hasn't stopped cities like Phoenix from blooming in the desert Southwest—they simply drain some distant watershed to quench their thirst. At least here there's no critical mass, no pyramid scheme of development to draw human beings into the void. Just two or three families, herds of cattle, horses, wildlife, grazing land, open skies—all you could ask for if you have a frontier spirit.

May 3, 2000—I created some maps of the ranch on my computer, based on a copy Sam gave me of the "Guy Cowden Range," and I'll make notes on them while I'm here. Sam and Herman, of course, need nothing on paper to know exactly where they are. You could knock them out and stash them someplace, and they'd wake with immediate recognition of their surroundings. I'm getting to know the ranch better with each successive visit and from research, but I've still got lots of country to cover.

The ranch is an irregular pentagon, thirty-six miles in circumference, its west line seven miles long, a south line almost ten miles, a short southeast line about three miles, a northeast line seven and a half miles, and a crooked northwest line eight miles. The quirky shape comes about because the land was part of a nineteenth-century Mexican land grant—the Anton Chico Grant, which formed the basis of later land sales. At the most northern and eastern points of the ranch are reliable springs; the original grant probably zigzagged to include them. With various Indian, Spanish, Mexican, and American claims, land and water in New Mexico have a long, complex history and a tangled legal trail. I'll have to track down the particulars for the ranch.

According to the map, the Guy Cowden Range contains 50,153 acres, although in the West, working ranches are often measured in sections, each one 640 acres. That's means the Cowden Ranch includes more than 78 square-mile sections. Headquarters is located in the southern part of the ranch, in Guadalupe County; but more than half of the land, on the north, is in San Miguel County. Since the ranch straddles the line, distant from both county seats, it's less subject to benefits or controls from either county. The road from town, for example, is a Guadalupe County road, only fitfully maintained—it's frequently rutted and rough. Resources in the county are modest and directed toward greater numbers of residents, rather than to a ranch at the end of the road. Political considerations also come into play, since the ranch represents only four votes—Sam and Kathy Cowden, and Herman and Debbie Martinez. Most county residents are of Spanish ancestry; Anglo ranchers like the Cowdens may own lots of land but have limited political influence. On a broader scale, lightly populated western states, with economies still based on natural resources, often take a colonial attitude: part detachment, part resistance to more powerful states and regions. Says Sam—independent, self-sufficient, leery of government— "We're outsiders."

More important than its place on the political map is how the ranch lies in the landscape. A mesa bluff runs from the northwest toward the southeast, dividing the ranch into two distinct topographies that affect how Sam uses it to run cattle. Up on top, the mesa gradually slopes back north toward the Conchas River; on the lower side are the headwaters of both Alamito and Esteros creeks, which feed the Pecos River to the south. Alamito drains the western edge of the ranch, occasionally carrying runoff toward the Gallinas River and thence to the Pecos. You'd never know, however, that the "creek" on maps exists in the world; Alamito appears as an eroded arroyo, almost always dry. Esteros Creek also flows only intermittently, but it's been a historically significant feature, as the ruins of rock houses and corrals, and the reports of various explorers, testify. Crossing the Llano Estacado from the east, or ascending the Pecos River from the south, Esteros is the first creek heading in the Rocky Mountains to the northwest. Most travelers were bound for Santa Fe, seventy-five miles from the ranch, and Esteros was an important sign of progress toward that destination—one party of Anglo explorers dubbed it "Hurrah Creek."

Esteros waters the more protected southern pastures of the ranch, where Sam keeps the bulk of his cows. The Cowden Ranch is a cow-calf operation, with a permanent herd of breeding cows producing a calf crop each year. Calves are born early in the spring and weaned in the fall, when they are either sold or held over through the next year. Other ranchers might buy and sell cattle every year, producing none themselves; such "stocker" operations are flexible, especially for where it's difficult to winter livestock or for rangeland subject to drought. Sam also runs yearling steers and heifers, most of which he raises from his own cows, but Cowdens have always been cow-calf ranchers. A cow herd is a living foundation, just like land; generations of cattle and cattlemen are behind Sam's calf crop each year. That's hard to duplicate at a sale barn. Some ranchers are, at heart and in practice, cattle buyers. Not Sam. Buying cattle is risky; one never knows what the market's going to do, or the weather. Those things affect him too, but he doesn't have to lay out cash each year to stock his range. His life is *raising* cattle, which goes hand in hand with the family tradition here. "It's what we've always done," Sam says. "It's what I know."

As a cow-calf man, Sam needs to have protected pastures for his mama cows; the lower side of the ranch is ideal with its bluffs and creek bottoms. On the open, mesa-top pastures, Sam keeps his yearling cattle. Access to working pens is also a consideration, since cows and calves require more attention. Sam's most complete pens are on the lower side—the Shipping Pens near headquarters and the pens at Alamito. A smaller set of working pens is located up on the mesa, and there are other small, rarely used pens in the southwest and east corners of the ranch.

The ranch as a whole is situated on the cusp of the southern Great Plains and the southern Rocky Mountains; except in spots, the land is neither rugged nor featureless, as it is to the west and the east. There's lots to say about climate and water, but after fifty years in the family, each generation making improvements, the Cowden Ranch is reasonably well watered—if drought conditions don't exist, and they frequently can.

Pastures have surface tanks, windmills, springs, or pipelines that supply water for the cattle. The basic grass is blue grama, with other native grasses like buffalo grass and galleta available for grazing. "Short grass," my grandfather used to say, "but strong. Cattle do well on it." Cholla cactus, yucca, mesquite, and other vegetation compete with grasses; junipers and piñon pines dot the rocky hillsides and tops of mesas.

There are ten major pastures, as well as two horse pastures and assorted "traps" used primarily for ease of moving, holding, or working cattle. When Guy Cowden bought the ranch fifty years ago, there were two or three enormous pastures. Now the largest pasture is Eversaw, almost eight thousand acres; three others contain more than seven thousand acres each; the remainder run one to three thousand acres. The scale of the ranch impresses even stockmen from back East, where entire farms would fit into a single "small" pasture here. But it takes lots of land in dry country to run cattle—twenty to fifty acres for each cow and calf. Back on our farm in Mississippi, with intensive pasture management, we used to run one cow/calf to each acre, rotating herds around thirty- to fifty-acre pastures. I always felt that the smaller pastures and the tractor work in Mississippi were an indication of illegitimacy—real ranches and real cowboys needed more room. That attitude arose in part from pervasive romantic visions of the Old West and in part from my actual experiences in Texas and New Mexico, where my first impressions were formed.

Barbed-wire exterior and cross fences are in place, some of them aging, few of them new. Sam uses electric fences in only a couple of spots; fence lines are too long for them to be reliable. There are close to one hundred miles of fence to maintain, and more than a hundred miles of dirt roads snake around the ranch, sticky or stony in places, graded as necessary or as time permits. The ranch has its own road grader, a great aging yellow beast that Herman and Sam struggle to keep running.

Land, water, fences, pens, roads, houses and barns, vehicles, livestock—the catalog of capital investments is staggering, and one reason fewer ranchers operate every year. What's enabled Sam to stay in business is partly the long history of ranching in the family; the efforts of previous generations continue within his efforts today. Decade after decade, the Cowdens improved their ranches as time and resources permitted, passing along the benefits: William Hamby to Gene to Guy to Rooster to Sam. What galls many failed ranchers, perhaps more than their lost livelihood, is the loss of their heritage. You don't fail yourself only; you fail ancestors and descendants. "It might end with me," Sam says. "These kids might not want any part of it. Or maybe it won't be possible anymore to make a living ranching. I hate to think it, but it might end with me."

🌾 🌾 🌾

May 4, 2000—A quick cowboy glossary, addressing the three principals of ranching: cattle, horses, and men.

Cows are most specifically female cattle, after they've had a calf; female calves are called **heifers** before they reach breeding age and **first-calf heifers** until they wean their first offspring, usually at about three years old. **Steers** are castrated male cattle. Among the different breeds of cattle are **Longhorns,** of Spanish origin, and British **Herefords** and **Angus,** the latter a hornless breed. Hereford-Angus crossbred cattle are called **black baldies** or **black whitefaces.**

Horses are most specifically males. **Mares** are females; **geldings,** castrated males; **stallions,** breeding males. A young female is a **filly;** a young male is a **colt.** Breeds include **quarter horses, Thoroughbreds,** and many others; **mustangs** are wild horses of Spanish ancestry. Horses are frequently referred to by their color, such as **bay, dun, palomino, pinto, sorrel,** and so on.

Cowboys are hired **hands,** also called **cowpokes** or **cowpunchers** (from those who worked in rail or stockyards, moving cattle along chutes); an old-time **cowhand** was called a **waddy.** A **cowman** raises or runs his own cattle; he doesn't work for wages. Cowmen are also called **cattlemen** or **ranchers.**

All these categories have many variations, and the ranching world is rich with lingo derived from Spanish and Anglo roots, from the physical landscape, and from practices of the range. When a waddy snakes a hoolihan around a snuffy, slaps his tree on, grabs the apple, and hazes some corrientes toward a democrat pasture, you're seeing pretty punchy stuff. No savvy? Rustle up a copy of *Western Words,* by Ramon Adams.

🌾 🌾 🌾

May 5, 2000—Dawn on Cinco de Mayo, a Mexican national holiday and Sam Cowden's birthday. On May 5, 1862, outnumbered Mexican troops defeated French forces at Puebla, ending Napoléon III's ambitions in Mexico. Cinco de Mayo is celebrated now on both sides of the border; go to any Spanish-speaking area in the Southwest and you'll find a fiesta. Sam likes to think of it as a holiday in *his* honor, so I've got a Cinco de Mayo T-shirt to give him later today.

And May 5 was the day in 1541, say some historians, that Coronado and his army had to delay their journey toward mythic cities of gold because the Pecos River was flooding with snowmelt and they had to build a bridge to cross it. Some believe they crossed near La Junta, the junction of the Gallinas River with the Pecos, placing them just ten miles—one day's march—from where this ranch house now stands. Ten miles, four hundred and fifty years.

This is Sam's day, and it might seem like Sam's ranch, but there's more to it than one man. A cowboy might make it alone, just him and his horse, but most cattlemen depend on many others. Ranch life—rooted in place, isolated, and labor-intensive—calls for a family. Men may make ranches work day by day; women make them survive and flourish over time. Sam knows he's lucky to have Kathy: "I didn't know if I'd ever find anyone to come way out here."

Kathy grew up in Ballinger, Texas, where her father has a ranch. I say "father" rather than "family," since everyone—Ralph Spreen; his wife, Joyce; Kathy; her brother, Kevin—lived in town; Ralph drove to the ranch every day to tend his business. "He's got real good Hereford cows, also sheep, and he works hard," says Sam. "I mean *hard*. Six to dark, every day, day in, day out. Make it work when something's not right; work harder at it. German." Since the days of founder Stephen Austin, Texas has seen waves of German immigration; places like Fredericksburg and New Braunfels and Ballinger remain influenced by their German heritage, with disciplined, industrious Volk. Kathy never seems to rest, among her roles as mother, teacher, ranch wife. Sam himself works hard, but he's not dogged. Sometimes Kathy can't fathom why Sam's schedule keeps changing. "Drives me crazy," she says. But cattle work in an open environment requires patience and flexibility; Sam might be boss, *el jefe*, but he can't call all the shots.

More to Sam's point, however, is the division between "Ralph's work" and "Ralph's ranch" and his family. Kathy and Kevin helped out when needed but apparently never felt themselves truly involved in the enterprise. Sam would like to shift Kathy's attitude toward the "family ranch" so everyone, especially their children, will take it to heart. "This isn't *my* ranch," he says. "It's *ours*." Kathy believes this even when she compartmentalizes their roles—necessarily, since her priority must be the homeschooling she provides for the kids. Schooling is part of the whole enterprise, as Kathy and Sam realize, an alternative to living in town or sending children off to school. So Kathy tends to give Sam responsibility and credit for the ranch and focuses her energies elsewhere. She's also self-effacing by nature, although no one lasts long in this gritty world without grit.

Before Sam and Kathy came Rooster and Jean Cowden. "I was a city girl," Jean says. "No one thought I'd stick." She did, of course, though since Rooster's death in 1994, she has lived in Santa Rosa. Jean lets Sam run the ranch, just as she deferred to her husband. Sam says, "Daddy was always the only word. What he said, by God, went. Now he's gone, she's found out she likes to have her say." But Jean says, "Sam doesn't ask *me*. Been on the ranch fifty years, what do I know?" Sam actually seems to like Jean's newfound independence, and she doesn't really want to run the show. There's good humor rather than animosity in their observations of long and close family ties.

Sam's sisters, Christy and Patty, remain partners in the Cowden Land and Cattle Company, which owns the ranch. For five generations Cowden siblings have raised cattle and held land in common, with remarkable harmony. Unlike some famous feuding families, they seem to have escaped the infighting that good fortune can produce. There have always been those willing to carry on ranch work and those willing to step aside—generally following patrilineal traditions—but no one is excluded. Ranching is in everyone's blood. Christy Cowden Brown now lives in Eldorado, Texas, with ranching husband Matt and their children; Patty and her husband, Souli Shanklin, live with their three sons near Rocksprings, Texas, where they have a ranch. Souli and his brother, Jake, often come out to help Sam during branding season, bringing horses they are training for cattle work or as polo ponies. Souli is due tomorrow, his tenth straight year. Because of their families and school schedules, Christy and Patty don't come, although both are excellent riders, better than I'll ever hope to be. In any case, they remain involved in the ranch where they grew up.

So, despite the masculine character of day-to-day cattle work, ranching for the Cowden family includes Sam, Kathy, Jean, Christy, and Patty: one man, four women. The next generation includes two girls, Abby and Hannah, and one boy, Guy Coley. This world might favor men, but the numbers are with the women. That was true in the previous generation as well; my mother was the eldest, followed by her brother and sister. And although my uncle Guy Tom always seemed like the quintessential cowboy—strong, silent, manly—I learned that when they were little children on the ranch, my mother put a dress on him and had herself a playmate in "Guy Margaret." And years later he was sent to fetch my mother, who had traveled from Midland to Roswell, New Mexico, for the weekend. Disgusted but doomed, Rooster complained, "I'm nothing but a woman-hauler!" Cowden women have always been forces within the family.

I think that a deep seed for this book was planted when I returned to New Mexico for Rooster's funeral. I remember in particular the scene at the cemetery, where a cold wind was blowing, and Jean, Christy, Patty, and Sam sat wrapped in blankets. I was struck by the quality of the ranchers and cowboys from around Santa Rosa who came for the service and joined the procession from church to cemetery, driving pickups, many pulling stock trailers. All around the gravesite I saw the figures and faces of working cowboys, absolutely authentic in appearance and bearing. I suddenly realized, despite the date and the occasion, that the Old West was not completely lost. Far from Santa Rosa, the world rushed forward—civilization faster and more complex by the moment—yet these men and women seemed somehow untouched by all that. I know my epiphany was more emotional than rational, influenced by the mood of that time and place, but something eternal seemed present. Evident in the bowed heads of cowboys holding their hats—weathered faces, tops of their heads glowing white as halos—was an enduring way of life. Afterwards they came to speak briefly to the family before heading back to their pickups and trailers, spurs ringing as they walked away, to disappear once again on distant ranches. That image of archetypal working cowboys has never left me.

Among fellow pallbearers that day was Herman Martinez. I've known Herman since we were boys, chopping cockleburs out on the ranch. He often came out from town with his father, eventually moved to the ranch to work full-time, and has become an indispensable part of the ranch. And there was Leonard Lujan, who worked for Rooster, then Sam, for twenty-seven years. The ranch is remarkable for its stability—testament to both cattlemen and cowboys—but there are times when all ranchers need additional help. Now, for example. Sam left this morning at 3:30 to help at a neighbor's branding. In the old days, ranches sent around representatives both to help and to watch out for their own interests, since cattle of different brands mixed on the open range. "Reps" might be one trusted hand who worked beyond the ranch's usual range—an "outside man"—or a whole crew and chuck wagon sent to a general roundup. The practice survives as part of the contemporary code of the West, neighbor helping neighbor in a sparsely settled land. Sam had to drive halfway to Albuquerque and won't be done until noon. "Sure you don't want to rep the Lazy Six?" Sam asked. I'm no fool. I'll help Herman around here—move some bulls, do some feeding, light work. Tomorrow we'll gather a few cattle in South Sabino pasture, getting ready for the work next week. Sam plans to gather Monday, Wednesday, Friday, and brand Tuesday, Thursday, Saturday.

In addition to Herman, Juan Estrada has worked here full-time since last spring. Juan and his wife are from Mexico and came up on a temporary basis. "He's worked out fine," says Sam. "I thought he'd probably stay only one spring and summer, but he did so good, I kept him on." Sam's got other hands coming from far and wide. Souli is driving up from Rocksprings; Jerry Young is coming down from Colorado for the week; and I'm here from Massachusetts—a real stretch. Local cowboys are coming on days we will actually brand.

Last night, Mark Wheeler and his sidekick, Toby Foote, dropped in on their way to a local ranch. Anything within fifty miles is local. Sam gathered cattle with them last week up on the Canadian River and enjoyed riding with cowboys who really knew their work, performing at the peak of their powers. It was also nice, he said, not to be in charge. Mark and Toby were welcome here, of course; in the past, ranches traditionally kept coffee and a pot of beans simmering for travelers who might appear. Any wariness of strangers in an isolated place was balanced by the hospitality required by that same isolation. Front doors had latch strings on the outside, functional and symbolic both: *come on in.* I'm always struck by the contrast between the crowded, inhospitable Northeast and the Southwest. The lesson seems to be that increased exposure to fellow man produces distrust: *familiarity breeds contempt.* Which sounds sadly cynical in these particular circumstances.

We sat outside the bunkhouse, drank a beer or two, and visited. Mark is thirty or so, with a wife and three kids he hauls around as he goes wherever the work and country seem promising. His great-great-grandfather was a Texas Ranger somewhere near the Panhandle; Mark's carrying on the cowboy tradition. "But not," as he remarked about many hands now on the Bar Y, "like some buckaroo." Cowboys from the Northwest, from Oregon and Idaho and Nevada, buckaroos are different from cowboys from Texas, New Mexico, or Arizona—those Mark calls "southern cowboys." "Buckaroo" is a corruption of the Spanish *vaquero,* and the high style of buckaroos derives from old Spanish equipment and practices, particularly in California. Buckaroos favor lots of silver on their saddles and bridles, dress elaborately, and resist any work afoot. I'll be interested to hear more about this when Mark comes back to help brand next week.

Sam says Toby is a first-rate cowboy. He's eighteen, lanky, quiet; only after Kathy left to put the kids to bed did he begin to speak—so fucking profanely that I understood his reticence with a woman around. Many elements of western manners, including deferential treatment of women, derive from the southern origins of early cowboys. Despite the long presence of Spanish and Mexican settlers along the border, most of Texas and eastern New Mexico was actually settled by Anglo-Saxon stock from the southern United States, like the Cowdens. They brought southern culture, which mixed with Hispanic manners, just as southern cattle-raising methods combined with ranching practices from below the Rio Grande. The herd dogs, for example, that Sam and Mark talked about—each of them has Australian shepherds—were used originally in the Carolinas and other southern pine-barren states, where thick cover made dogs a necessary asset to herders on foot. Out here on the plains, cowboys on horseback have supplanted Carolina herdsmen afoot with cow dogs.

One problem is getting a full, competent, and healthy crew together. "People have bad knees, wooden legs, fused bones, whatever," says Sam. "Jerry's sixty or more. Given the physical demands, cowboys should be twenty, not forty or fifty or sixty. But this country is short on young cowboys." What holds the operation together is the unwillingness of individuals to fail one another and their employer. "Riding for the brand" is the old-time expression for cowboy loyalty and sacrifice—commitment to their outfit at all costs. At one time that involved niceties like gun battles. Billy the Kid used to roam this country; his reputation was made riding for the brand during the Lincoln County War, when rival ranchers fought for control of territory. Legend has it he once hid out here on the ranch, at Sabino Springs. He's buried just down the road in Fort Sumner, where the Billy the Kid Museum offers up crackled photographs, antique pistols, a stuffed two-headed calf, and such. Our job here is less complicated than Billy's—burn Sam's Lazy 6 on five hundred calves.

Last night after Mark and Toby left, I settled in. I'm sleeping on the porch of the bunkhouse, on a cot that Sam says was Rooster's favorite bunk. Souli and Jerry come tomorrow, each of them with seniority at these brandings, so they get the two bedrooms. I'm satisfied. The porch is cool at night, and I can listen to the coyotes yip in the distance. Anyway, I ride for the brand.

May 6, 2000—Today is Tommy's birthday. Sam's older brother, Tommy Cowden was killed by lightning here on the ranch when he was eighteen. "We were up on the mesa at the roping pens one evening," Sam says. "All of us—Tommy, Christy, Patty, Herman, me—on horseback. A cloud had come up back north, and Momma and Daddy went to check it out. When we were kids, lots of times we'd see the top of a thunderhead appear back north or west and everyone would jump into the car and we'd drive up on the mesa for a look. Often, by the time we got there, it would be gone. 'Whoops,' someone would say, 'we chased it away.' That's what Momma and Daddy were doing—chasing clouds."

The unobstructed skies of New Mexico are a source of constant wonder, gaudy at sunset, though often clear blue until afternoon, when clouds build above the mountains and drift out over the plains, with horizontal landscapes and vertical cloud forms intersecting to create unforgettable images. Willa Cather wrote in *Death Comes for the Archbishop*, "Every mesa was duplicated by a cloud mesa, like a reflection, which lay motionless above it or moved slowly up from behind it. . . . The great tables of granite set down in an empty plain were inconceivable without their attendant clouds." I'm also especially partial to "walking rain," where distant clouds spill gray veils of rain that never reach the ground. Called virgas by meteorologists, from the Latin for "stick," walking rain evaporates before it can dampen the dry landscape. And of course you see thunderheads, majestic cumulonimbus clouds rising thirty to sixty thousand feet into the atmosphere, where tops flatten into anvil shapes, sunlit in the evening long after the sun actually sets. However beautiful and welcome for their downpours, such clouds are powerful and dangerous weather systems.

On average, more than twenty-one million cloud-to-ground lightning strikes from thunderheads occur annually in the continental United States. Half produce more than one strike point, for a total topping thirty million lightning bolts. Each takes less than half a second. They produce an average eighty-seven fatalities each year, second only to floods in weather-related deaths. The highest death rates occur in July, between two and six in the afternoon, on a Sunday, among males (84 percent), in an open field (26.8 percent, twice the rate for any other location). New Mexico, despite being a dry and lightly populated state, ranks first in yearly lightning deaths per capita, 1.88 for each million people, or just over two deaths each year. One afternoon in June 1973, Tommy Cowden became one of those statistics. To his family, of course, the loss was personal, immediate, and lasting.

"There were clouds around," says Sam, "but nothing directly over us at the pens," located atop the mesa near headquarters. From there you can see for miles and miles, south and east toward Santa Rosa and the Llano

Estacado, north and west toward the Rockies. The Cowden kids and Herman used to rope calves all the time up there; now the pens stand empty and unused, a reminder of that fatal day. "It never rained, hardly a sprinkle fell," Sam recalls. "It was almost a bolt out of the blue. One moment we're riding along, the next . . . we're on the ground." The lightning bolt hit Tommy, burned a hole through his saddle, and killed his horse. Everyone else was knocked to the ground. The smell of charged electricity hung in the air as they scrambled up. They attempted to revive Tommy, couldn't, put him in a pickup, and raced toward the doctor in town, thirty miles away over dirt roads. Tommy never came back.

I remember attending the funeral, Tommy lying in a casket at the Chavez Funeral Home across the road from the Cowdens' house in Santa Rosa, and the always unreal quality of the dead. And I remember speaking with his father at the kitchen table in town. "He was just getting to the point where he understood the ranch," Rooster told me, stoical as always but certainly shaken by the loss of his eldest son, the child most likely to assume the reins of the ranch one day, just as Rooster had assumed them from his father. Although the family had always treated male and female children as equal inheritors, it was generally expected that a son rather than a daughter would remain to run the ranch. "He was just coming along," said Rooster. "Maybe if I had been there, I could have done something."

Rooster's sense of guilt was natural, one facet of grief, but what could he have done? Tommy was struck down on a ranch miles from medical care, with no telephone to call for help, no helicopter ambulances, no resources other than the wits and grit of his family. Ranchers have faced such conditions forever. An 1889 letter from Cowden relative Jerry Jowell describes a close call out on the JAL Ranch, when his sixteen-year-old son, Spence, took a hard fall from a horse while riding with Rorie and Gene Cowden: "Darling Wife . . . Spencer came very near getting killed on Friday after we got Home. Rorie & Eugene was with him—he was riding old Stray Gray and trying to rope a yearling. I Suppose just for fun. And the horse fell with him and then dragged him about 75 yards by his foot hanging in a rope—the fall knocked him senseless. The boys were scared so bad they could not take him to the House[;] they were out near the big gate. Gene went after Walter and Rorie built a Shade over him. When Walter & Cal came they thought he was about dead, he had never spoke[,] only had asked where he was—it was sometime after they began working with him before he came to his senses—he is up and about now— his shoulder right sore yet—his face is badly bruised. Some of his teeth were loosened and he thinks his nose about broken—the white of his right eye is as red as a ball of fire—badly bloodshot. I think he is all right now as soon as the soreness leaves him."

And Benjamin Harper remembers an incident while working west of the JALs: "In 1884 or '85 I went to Lincoln county, New Mexico, where I was range boss for Lish and Green Estes, who branded 727. Midland, Texas, two hundred miles east, was our nearest town of any consequence. . . . A rider had a horse fall on him and dislocate his spine so that he couldn't move any part of his body but his eyelids; couldn't even talk. We loaded him into a hack to which I hitched a span of half-breed mules and we covered the two hundred miles to Midland in less than twenty hours. Doc popped the man's bones back where they belonged and we returned to the ranch the next day."

It's commonplace for ranching folk to minimize injury and illness, part of their basic makeup, part of their condition. Self-sufficient and far from town, they work through minor ailments because they have to, and they eventually take some pride in withstanding pain of any sort. Rodeo cowboys with broken ribs nevertheless keep riding bulls, not only to win their event but also to show how tough they are. Without a peep, Souli hobbles around these days from old rodeo or barroom injuries. "Goes with the territory," my grandfather used to say about such troubles. The territory is tough and toughens everyone who sticks with it.

It's been nearly thirty years since Tommy's death, and the family has moved forward. It was an accident, after all, an act of God, and death too "goes with the territory."

May 7, 2000—Tommy's death reemphasizes how thoroughly this life is lived in the "natural world," a phrase that often implies a division I don't accept: that humans are not natural. It's the traditional conflict that literary critics hunt for: Man versus Nature. As if over time we somehow ascended from the ground of our creation into another sphere of existence entirely. As if "civilization" were removed from physical life, rather than a unique mode of it. Clearly, cattle ranching is not an enterprise of the office cubicle, the skyscraper that blocks out the sky, human constructions that defy the forces of nature. Less evident are the actual details—physical, emotional, spiritual, and intellectual effects upon those who live their lives in the open. There's a maxim of medieval warfare: "There is no reply, in open ground, to an archer under cover." Here in the natural world, archer and cover are the dome of heaven itself—cloud cover or fierce sunlight or relentless wind, to say nothing of the occasional antagonism of solid earth.

Last week I took a trip from the ranch over to Amarillo, Canyon, and Lubbock—Texas cities on the far side of the Llano Estacado—to visit research libraries and museums. I drove east from Santa Rosa to Tucumcari, Interstate 40 following old Route 66, the legendary highway that ran parallel to the Canadian River ten to twenty miles north. Spanish land titles in New Mexico referred to the Canadian as the Rio Rojo and the Rio Canadian, and some locals still call the Canadian the Red, which causes no end of confusion with the real Red River over in Texas. Just as red rivers repeat in the Southwest, routes repeat over time, and I drove the same path that Coronado and his army followed 460 years ago. After skirting mesas and the forbidding face of the Llano Estacado escarpment, beyond Tucumcari Peak the Spaniards ascended onto the plains, into the almost unbelievable wide open. As I drove, I tried to imagine the Llano as it was then, sun-shot and windswept, without the trees now grown up around farmhouses, without a single human structure, without any landmarks whatsoever. Only grassland stretching toward every identical horizon. The sea, so often used as a metaphor—the long ocean voyage from Spain to Mexico was still fresh—offers the same sights and sensations: open sky, wind moving over the swells, nothing to break its run across the waves, across the grass tops. The Spaniards looked out ahead and set a trackless path toward cities of gold. Both then and now, which is more unbelievable? The imagined or actual experience? Although the Spaniards had seen plains in their native León or Castilla, and stretches of desert below the Rio Grande in New Spain, they were inadequately prepared for the Llano Estacado. They found no relief on the plains, none at all, until reaching the eastern edge of the Llano, where they encountered, not mountains, but canyons descending into the earth. That's where I headed first, to Palo Duro Canyon.

Historians debate whether Palo Duro or some other canyon was one barranca mentioned in chronicles of Coronado's route across the southern plains. Those barrancas (from the Spanish for "ravine" or "gorge") were and are "so concealed that one does not see them until he is at their edge." Hours of driving over open country brought me, suddenly—even though I'd read of it, followed signs, anticipated it—to Palo Duro. The earth opened down and out a hundred miles into the hazy distance, a stream far below winding through stands of cottonwoods and cedars on the green canyon floor. Above loomed craggy bluffs, red ribbons of earth, fantastic pillars of eroded rock. The view was majestic, in the way we usually feel about mountains rising above us. But a *canyon*, a mountain in reverse, with the same elements found in mountains: water, vegetation, animal life, human history. Once I twisted down the road into Palo Duro, I could see why Comanches and other Indians had their faint trails across the plains converge here, why Charles Goodnight made a ranch of it, why later Anglos—the state of Texas—took it for a park. In the gathering shadows of late afternoon, among the shade trees and pockets of meadow grass along the shallow moving water of the Prairie Dog Town Fork of the Red River, the canyon cooled. I spotted wild turkey, deer, rabbits, birds, and people there to picnic or hike or camp. Arterial, fundamental, mysterious in how they brought forth life, river and canyon were an indispensable and rare resource—I had descended into a protected treasury of the plains, into a secret garden.

The protection of the canyon can't be overstated, especially compared with the exposure travelers faced on the open Llano above. After I left Palo Duro, at the Panhandle-Plains Historical Museum in Canyon, I discovered Coronado's army had suffered through one of the frequent violent hailstorms that sweep over the plains. From late spring through summer, warm moist air from the Gulf of Mexico flows up over West Texas and eastern New Mexico and collides with weak cold fronts moving off the Rocky Mountains. Cumulonimbus clouds build into towering thunderheads miles high—up to sixty thousand feet. Nothing in the sky or on earth impedes your vision of these rolling, bruising cloud forms, their tops like enormous anvils ringing down lightning bolts and thunderclaps. Also within thunderheads, hail can develop as air currents send particles rising and falling and rising within the cloud form, ice building with each ascent until it finally falls as hailstones. The most hail-prone region in North America is the Great Plains. The largest recorded hailstone in the United States fell in Coffeyville, Kansas, in 1970, seventeen and a half inches around—the size of a big cantaloupe—and weighing nearly two pounds. More commonly, golf ball- and baseball-size hailstones fall, sometimes accumulating two feet deep. Hail routinely destroys crops and damages roofs and cars and sometimes can kill. The National Weather Service reports that on May 13, 1939, near Lubbock, Texas, a farmer was killed when caught in the open during a severe hailstorm. Every cowboy learns to dismount in a hailstorm, unsaddle his horse, and cover up as best he can with his saddle.

For centuries, travelers and residents here have come to recognize the perils of "open ground," of the plains and their violent weather. Coronado's army was camped in the relatively protected confines of Palo Duro Canyon (or the adjacent Tule Canyon) when a hailstorm struck. The chronicle of

Pedro de Castañeda of Náxera recounts: "While the army was resting in this ravine we have mentioned, a violent whirlwind arose one afternoon. It began to hail, and in a short time such an amount of hailstones fell, as large as bowls and even larger, and as thick as rain drops, that in places they covered the ground to a depth of two and three and even more spans. One abandoned his horse, what I mean is that there was not a horse that did not get loose except two or three which were held by negroes covered with helmets and shields. All the others were swept away until they ran into the barranca. Some climbed to places from which they were brought down with great difficulty. If the storm which caught them in the barranca had found them on the level plain the army would have been in great danger of losing its horses, for many would never have been recovered."

Much the same thing happened this year with Sam Cowden's horses, which stampeded one night during a lightning storm, running panicked in the dark through a fence. Several horses were cut up, one—Gopher— so severely it had to be destroyed. Weeks later Herman was still tending Streak, whose chest and forelegs bore deep scars from the barbed wire— although you might equally blame lightning or hail from heaven. There is no reply, in open ground . . . You duck and suffer your fate. Here you are a part of the natural world, not apart from it.

May 8, 2000—We will gather Bogie Pasture today and brand tomorrow at the shipping pens. No local cowboys are coming for the gather, but we have plenty of riders: Sam, Herman, Juan, Souli, Jerry, and me. We'll start right after breakfast, and Sam expects us to finish before noon. In the other room, the television is on fire. Near Los Alamos, forty-seven thousand acres of forest have burned—the largest fire in New Mexico history, almost the size of Sam's entire ranch. They've evacuated the city, and the Los Alamos National Laboratory, with its stores of nuclear materials, is in danger. We can see smoke from the fire—a hundred miles northwest—from here. Crews are fighting other fires at Pecos and Las Vegas, and farther south in Ruidoso and Cloudcroft, where twenty thousand acres have burned. This may be the worst fire season on record in New Mexico, due to dry conditions, high temperatures, and strong winds that feed the fires. La Niña seems to be responsible for even warmer and drier weather than usual; Pacific Ocean weather somehow helps produce severe to extreme drought conditions here. "It's been dry for two years," says Sam. In place since late 1998, La Niña has been a bitch in the Southwest.

Along with the fires, arguments rage about policies. "Fight all fires" is no longer the rule but was in effect for most of the twentieth century, producing an enormous fuel load of small trees and brush, rather than allowing periodic low-intensity fires to thin forest undergrowth. The Los Alamos fire came from a prescribed burn, designed to reduce the fuel load and threat of a high-intensity, high-speed fire. Yet that's exactly what they got. Set by U.S. Forest Service personnel despite drought conditions and gusty winds, the fire quickly burned out of control and all hell broke loose—more evidence of dicey human involvement with the forces of nature.

Right here, Sam had a prairie fire sweep across part of his ranch back in March. It started on the other side of the Gallinas River at Park Springs, John Brittingham's ranch, where burning trash got out of control. High winds pushed the fire east, and it jumped the river and moved onto Sam's ranch. Juan called from Alamito about noon, "Hay fuego aquí!" ("There's a fire here!") Sam and Herman raced over and were joined by John Brittingham and eventually by other neighbors, the fire departments from Santa Rosa and Anton Chico, and, finally, Forest Service personnel. They fought the fire all day, slapping at it with wet gunnysacks, trying to control its spread and progress. Grass in the pastures was short, so the fire stayed relatively low, although it moved fast, fanned by fifty-mile-per-hour winds. Sam's road grader was broken down, parked near Alamito; incredibly, the fire didn't burn it up, though it sat vulnerable—dripping diesel and oil and hydraulic fluid—in the pasture. On the other hand, four miles of Sam's fence burned up, old cedar posts turned into torches and strands of barbed wire ruined by the heat.

After burning five thousand acres, the fire was eventually contained where the mesa rises east of Alamito and the soil gets rocky. "There wasn't as much fuel for the fire," Sam says, "and several other things helped us. The wind dropped off just about dark, like it often does—'Can't see to blow,' they say—so the fire slowed down. At that same time the Forest Service guys arrived, and so did the road grader from the Bar Y. We were able to get fire breaks in place and get it stopped. I guess there were thirty of us fighting it by then.

"It's lucky we stopped it at the mesa. Had it reached the top, there was nothing to stop it. It would have burned all the way to Amarillo. Probably would have burned up headquarters, and you know we can't get insurance way out here. We could have lost it all.

"It was mighty strange out there near the end—night coming on, flames moving at us, wind blowing, tumbleweeds on fire rolling and flying by. Very spooky. Very, very spooky."

The gather this morning went fine. I've had a siesta and a chance to look back into my reading notes about prairie fires. A laptop is a wonderful thing.

The secretary of agriculture said yesterday on television, "We have the hottest, driest weather in perhaps fifty years; we have thousands of lightning strikes an hour; we have three hundred new fires every day in the West, largely because of lightning strikes." But others claim that only 10 percent of wildfires are the result of lightning and that humans start 90 percent of the fires, by accident or design, and have for centuries. Long before the federal government set policy, Plains Indians set prairie fires to drive game toward hunters or as a tactic of combat. Here's A. B. Gray, in his *Survey of a Route on the 32nd Parallel for the Texas Western Railroad, 1854:* "In burning the grass behind them, the object of the Indians is to accelerate the spring growth, and prevent immediate pursuers having forage for their animals. A war party had a few days previously passed along, and for many miles the fire was raging around us. As far as the eye could witness, all was a-blaze, and at night appeared like a vast amphitheater of illuminated cities. This is not one of the least beauties of the prairies. It was not unfrequent to find large isolated patches, untouched, through changes of the wind, upon which we encamped, suffering but little inconvenience for want of grass. A narrow space previously burned off, or an ordinary road, will often stop the progress of the flames."

I found two similar accounts in Josiah Gregg's 1844 *Commerce of the Prairies:* "The old grass of the valley in which we were encamped had not been burned off, and one of our cooks having unwittingly kindled a fire in the midst of it, it spread at once with wonderful rapidity; and a brisk wind springing up at the time, the flames were carried over the valley in spite of every effort we could make to check them."

And while east of the present Cowden Ranch, Gregg related: "We had just passed the Laguna Colorada . . . when our fire was carelessly permitted to communicate with the prairie grass. As there was a head-wind blowing at the time, we very soon got out of reach of the conflagration: but the next day, the wind having changed, the fire was again perceived in our rear approaching us at a very brisk pace. . . . The worst evil to be apprehended with those bound for Santa Fe is from the explosion of gunpowder, as a keg or two of twenty-five pounds each is usually to be found in every wagon.

When we saw the fire gaining so rapidly upon us, we had to use the whip very unsparingly; and it was only when the lurid flames were actually rolling upon the heels of our teams that we succeeded in reaching a spot of short-grass prairie, where there was no further danger to be apprehended. The headway of the conflagration was soon after checked by a small stream which traversed our route."

It's possible that the small stream Gregg mentions is Esteros Creek, with its headwaters on the Cowden Ranch, and it's certainly clear that Sam's prairie fire, however novel and difficult for him, was not unusual. Wagons with kegs of gunpowder become road graders full of diesel fuel, and fires are still "carelessly permitted to communicate with the prairie grass." And though it's hard to take out 10 percent of your productive grassland without preparation, Sam's burned pasture might come back better than before—if he gets adequate rainfall on the reseeded sections. Right now the grass is up and green, with charred remains of cactus rising here and there, like the chimneys of burned cabins.

May 9, 2000—This morning we branded the calves from Bogie, the first time I've seen Sam's method, which is very different from how we used to work our cattle in Mississippi. There we branded and vaccinated the calves in the squeeze chute; we castrated the bull calves in a small pen, flanking them on foot. No roping whatsoever. Out here they do things much the same way they were done on the open range in the nineteenth century, before there were working pens and squeeze chutes and calf cradles. A cowboy on horseback ropes a calf—ideally catching it by both heels—and drags it to the fire. Open-range branding fires were built of any available material, mesquite or other wood or cow chips, dried dung that made a stinky, though hot, fire on which irons were heated. Sam uses propane gas to heat his irons; the "branding pot" is a tube welded from heavy pipe that allows the irons to heat handily at waist level. Herman usually brands, rotating three irons so he's always got a red-hot tool to work with. Sam's Lazy 6 goes on the left cheek and left hip—ideal locations for identification of cattle throughout their lives. I remember as a child first smelling scorching hide in a branding pen, the odor carried on blue and yellow billows of acrid smoke that made me gag. What was nauseating then has become tolerable with experience, and burning hide now smells keen, almost sweet.

The cowboy dragging calves is at the center of the branding world—it's an honor to rope. You are the only man on horseback, performing a traditional skill, while others labor on the ground. At his own brandings,

Sam never ropes; code demands that the boss ask only visiting cowboys, some of whom travel hundreds of miles and work for free just to rope twenty or thirty calves. Roping is an art that has developed over centuries from its roots in Mexico, where vaqueros used *la reata*—Spanish for "rope," later Anglicized into "lariat"—to capture cattle. Early Texas cowboys adopted the equipment and methods of vaqueros on open-range ranches, roping cattle to brand or doctor them. At times a cowboy on horseback was alone in this work; other times a whole outfit would round up and brand many cattle, like we did today. The cowboy dragging calves in an open-range roundup had to identify the mother of each calf he roped and call out her brand so her calf would be similarly branded and marked: Lazy 6 or JAL or CC. Now the cowboy dragging just needs to keep the crew supplied with calves; missing loops slows down the whole operation and subjects the roper to commentary from waiting cowboys. The ultimate blunder is to drag up a calf that has already been branded.

Once a calf is roped, it's dragged by the heels to two men waiting on the ground—the flankers. Flanking calves is physically demanding. One man grabs the catch rope, the other grabs the calf's tail, and they jerk in opposite directions, flipping the calf onto its side. Timing is critical, like the violent choreography of a WWF match. One flanker quickly kneels on the calf's neck and holds a foreleg to prevent it from getting up; the other grabs one back leg and stretches out the other with his own legs. One moment, calf and flankers are standing; the next, all three are on the ground, locked in a rough, tussling embrace. Calves can range from a week or two old, perhaps 125 pounds, to three months old and nearly 300 pounds. A big calf can kick the shit out of you, and will if you let go, so you don't. Flanking calves for two to three hours leaves a man exhausted, sore, and usually bruised. It's work for the young and hearty; Sam kindly did not ask me to flank, although I did a time or two for the thrill.

Once a calf is flanked and on the ground, it's branded, vaccinated, and either castrated (males) or earmarked (females). The whole process takes about thirty seconds, an assault that leaves a calf bawling and somewhat stunned. Not for the dainty is branding cattle. Herman brands the cheek first, then moves back to the hip, trying to burn a clean Lazy 6 in the same place each time, hard to do if his flankers don't hold the calf steady. At the same time, another cowboy vaccinates the calf for blackleg and other diseases, one shot in the muscle, another under the skin, both delivered from pistol-grip syringes. At the same time, another cowboy castrates bull calves with his knife, cutting off the tip of the scrotum, reaching in, pulling and severing each testicle, then applying wound spray. It's important work for an experienced hand, since a poor job of castrating can kill a calf. Bull calves are also implanted at the base of one ear with a growth hormone. Female calves are earmarked, a wedge of their right ear clipped out—a mark called a V-underbit. If the calf has horns developing because of Hereford blood, those nubs—little plugs of relatively soft material—are dug out of its skull with a "horn spoon" and the wounds sprayed. That's all folks. The calf is released, to scramble up and wobble off, wondering what the hell just happened. Inhuman some would say, and they're right—these are cattle.

We had lots of hands available today—Ty Monisette drove in from his place up on the Canadian River—so there were two sets of flankers, one on each side of the fire, and calves were dragged to alternating sides. The cowboys branding, castrating, vaccinating, earmarking, implanting, and dehorning worked both sides of the fire. When there was a good roper working, the pace was brisk. That's what we all wanted, so we could finish before the day got hot—for the sake of cattle as much as cowboys. The aim is to reduce stress as much as possible, to do brutal work in a quick and quiet way.

I should have mentioned that, first thing in the pens, we "stripped" the calves, separating them from the cows. One man stands at a gate and lets cows pass through, but no calves. Small bunches are worked until the whole herd is sorted; cows are moved to a holding pen and calves into the branding pen, about half the calves at once. Too many and it's hard to throw a clean loop; too few and the calves dash around, also making roping difficult. After working the calves, we ran the cows down the chute and applied insect control down their backs, work that went fast and easy. Then cows and calves were reunited and slowly moved back into the pasture, where they could recover from the stress they'd just been through. We branded 140 calves before noon, with one break for snacks that Kathy brought for the cowboys. There were 77 heifer calves and 63 bull calves—now steers. For his final tally, Sam counts all the calves and figures males and females by the number of sacks saved from the castrated bull calves. We saved all the testicles to fry up later; "calf fries" or "Rocky Mountain oysters" are the special treat from a hard day of branding.

May 10, 2000—We gathered Eversaw this morning. "That came together well," Sam said as cowboys brought in cattle, arriving from different parts of the big pasture at about the same time. Souli rode the top of the mesa to look for any strays; I followed the lower rim, swinging along below him;

then we pushed our cattle toward the main bunch farther west. My horse picked up a spike of cholla, limping suddenly; lucky I had gloves to pull the thing off his ankle. Eversaw has lots of cactus and yucca that can really stick it to you.

Sam was courteous with the sort, giving Souli the opportunity to cut out some cows since he's been coming to help Sam for so long. As a horse trainer, Souli could always ride like nobody's business and over the years has come to work cattle well—the two skills don't necessarily go hand in hand. A classic book about the Bell Ranch east of here, Jack Culley's *Cattle, Horses, and Men,* invokes the holy trinity of ranching. Souli has now put all three together, like Sam, whose human skills I admire more and more. There was a moment when I was about to point out a dry cow, but Sam stopped me. "Souli will spot her," he said. "Unless he asks for help, we'll let him do the job." As always, Sam was careful to give a man a chance. Herman, Juan, Jerry, Mark, and I held the herd while Souli, then Sam, made the cuts. We were left with about 140 pairs of cows and calves.

This afternoon we had more bunkhouse discussion about buckaroos and ranching around here. Mark Wheeler and Ty Monisette have both worked for Henry Singleton, who owns all the land surrounding the Cowden Ranch—close to half a million acres. "He bought the Bar Y, Conchas, Latigo, and Payne ranches, everything around this place," says Sam. "Besides this division, he's got the San Cristobal Ranch over near Santa Fe and other ranches around the state. You hear talk about Ted Turner owning more land than anyone in New Mexico, but I think Singleton's got him beat. He's just quiet about it. Turner looks for headlines; Singleton looks for profit." Cofounder of Teledyne Corporation and former trustee of Ronald Reagan's blind trust when Reagan was president, Henry E. Singleton of Beverly Hills, California, has deep, deep pockets. He's been buying ranches in New Mexico since 1987, often at a discount when ranchers, like the Driggers family at the Bar Y, got into trouble. It's ironic that the Cowden Ranch, at fifty thousand acres one of the largest in the state, is dwarfed by its neighbor.

"Alex Carone is the ranch manager," Sam tells me. "Came from Nevada. He's done a good job turning the place around. I bet they didn't have a fifty-percent calf crop before he arrived, calves coming year-round, no management at all. When he came, a lot of Nevada buckaroos followed." Ty and Mark have their differences with buckaroos, who might make good cowboys on horseback, Mark says, "but they won't get down from the saddle to take a shit!" Who does the fencing, the ground work in the branding pens? "We did—southern cowboys," he says. "They're too good for that." Buckaroo style includes saddles dripping silver, tooled bridles, distinctive hats and chaps and spurs, and long ropes. "Why would a man carry a ninety-foot rope?" asks Sam. "He can't throw his loop over thirty feet. They've got ninety-foot ropes in sixty-foot branding pens!"

The answer to that one is partly historical, deriving from the Spanish and Mexican origins of buckaroo style. Originally la reata was woven of rawhide, which made a supple if not strong rope. To take the pressure off, vaqueros used long reatas and would play out rope until the beast at the other end grew tired, much like an angler playing a fish. Vaqueros would *dar la vuelta*—"give a turn" around the saddle horn—and use friction to slow or hold their quarry, releasing line if necessary; never did they tie their ropes to the horn. Over the years, *dar la vuelta* was anglicized into "dally," still practiced by many cowboys even though modern nylon ropes are far stronger than the original equipment. A rope dallied around the horn gives a cowboy more options and saves wear and tear on both cow and horse. Texas cowboys at first used Mexican techniques, only later tying stronger ropes "hard and fast" to the saddle horn. Thus long ropes— though not dallies—are an affectation, a kind of cowboy reenactment. "Southern cowboys are more practical than buckaroos," says Mark. "Down to earth."

Sam makes the comparison clear: "Yesterday our crew of ten cowboys branded one hundred and fifty calves in five hours, two men dragging calves. I worked a roundup at the Bar Y, and their crew of twenty-three cowboys, with five draggers, took thirteen hours to brand three hundred calves. Twice the calves, but more than twice the cowboys, and lots, lots slower." Mark shakes his head, adding, "They won't hustle."

The rivalry is new to me, and it's strange to think of Sam's ranch as a scrappy underdog to Singleton's operation. But Sam is in fact surrounded, the hole in the doughnut, the wagon train circled by Indians. And he's not even directly in competition with Singleton, since the cattle market remains an almost classic free market, supply and demand treating large and small alike. Or so I thought. It turns out that not long ago the New Mexico cattle market was skewed by Ted Turner, who converted his 588,000-acre Vermejo Park Ranch into a buffalo range, disrupting the market for cattlemen when he did. As Sam says, "It was a bad time. We were dry here, bad drought, and I'm sure they were dry at Vermejo too. Everybody was having to sell cows. There were probably, gosh, seven thousand cows on that ranch, and Turner said, 'I want them all off, and I want them dead.' They killed every one of them. There are only two cow-killing plants anywhere close. There's one in Hereford, Texas, the next is in San Antonio, and the next one going west is all the way out to Los Angeles. We couldn't sell our cows. And it hurt. Out in Arizona, a long way from everything, a lot of those cows just died. Ranchers couldn't sell them at any price. Ranchers in this part of the world just do not like Turner. He doesn't care."

If Ted Turner's nearly unlimited resources and indifference to the local economy make him a bad neighbor, that isn't true of Henry Singleton, according to Sam. "He cares about his ranches," Sam says. "He's not a nut like Turner. I may not like being surrounded, but he's never treated me unfairly. He's just mighty big."

Big private or investor-owned ranches are nothing new to the West. The XIT and its three million acres were controlled by a group from Chicago and London; Charles Goodnight was bankrolled by John Adair—thus the JA brand—whose money was from New York and England; the Matador, another famously large Texas Panhandle ranch, had Scottish investors. In the heyday of the open range, cattle ranching appealed to eastern money and foreign gentry; today media moguls and conglomerate billionaires play the game. Over the years, however, the Cowdens have always operated independent, family ranches on a more modest, if still large, scale. Although income from Texas oil royalties has helped ensure their survival, ranching has remained their principal work, not a hobby. Sam approaches it from the pasture up, not from a distance, and has five generations of ranching life behind his perspective. No one's going to mistake Ted Turner for a rancher, and no one can mistake Sam Cowden for anything else.

May 11, 2000—Today we worked the herd at Alamito, and it was hell. The wind's been blowing hard since dawn, steady at forty miles per hour, gusts to fifty. Big Guy used to say, "The wind won't quit until it rains; it won't rain until the wind quits." I asked Sam why the wind blows so much, so hard. "Nothing to stop it," he answered. True enough, but only part of the story; what starts it blowing? Is there no beginning, as well as no end, to the wind?

Wind can make horses jumpy and cattle hard to handle, tough to gather and drive. Sounds and scents scatter on the wind, so cows with calves grow nervous, particularly heading downwind. Wild cattle range into the wind, and a herd "pointed into the breeze steps out with an assurance and a right free will not altogether due to lessened heat, dust, and flies," Charles Goodnight observed. "A stampeding herd circles into the wind to know what is ahead. In working a roundup, cattle should always be cut into the wind—they work easier that way. And a cowboy, loping into the breeze, feels an exhilaration that stirs his blood like old wine." Not always. We got our cattle into the pens all right, but then the shit hit the fan. We set up the firepot on the west end of the branding pen, so dust would blow downwind from where we worked, yet there was no escaping a sand-blasting. The ropers threw windblown, messy loops at spooky calves and dragged them to us, horses and calves kicking up dust the wind sent ripping over us. We all wore sunglasses and pulled up our bandannas like outlaws, but everyone ate plenty of dirt. Miserable conditions.

Jerry Young had a double-bad day, fighting the elements and a terrific hangover. Last night the cowboys, Jerry in particular, got into a bottle of Lord Calvert Canadian whiskey that Sam had bought as a treat for later in the week, after the work was over. Jerry and Mark jumped the gun like Oklahoma sooners, and Señor Young got completely uncorked. By ten last night he'd fallen out of his chair, smashed his face, and suffered assorted other indignities before Souli dumped him into bed, fully dressed. About three o'clock this morning he was in the bathroom hurling and heaving, and only some bleary residue of pride got him into the saddle today. I took his place at the cookstove, making breakfast for the crew while Jerry commiserated with his horse out in the corral; no way could he face a griddle of bacon and eggs. And he knew better than to try roping calves, which would have been disaster. Soon he even had to give up vaccinating and retire to a corner of the corral, where he rested in a sorry heap, head down between his knees, while branded and castrated calves milled around him. I hadn't seen a drunk of that magnitude in a long time. Souli memorized each comical embarrassment, and now Jerry gets to relive everything he can't remember of last night.

The rest of the crew persevered, all of us looking like coal miners by the time we were done, caked with grime in addition to the usual gore, our eyes and ears and throats full of dust. What's more, this bunch of calves had gotten into patches of cockleburs and their tails were full of sharp burrs. The flankers wore gloves, but it was still one more level of hell.

Sam says the camp at Alamito always gets more wind, sitting out there in the open flats that lead to the Gallinas River. When he was growing up, they got much of their electricity from a wind charger. And Jean tells me that before she got a clothes dryer at Alamito, she would hang her laundry on the line, and by the time it dried, one side was clean and the other—the west side—was red from blowing dust. Which recalls one of the major discomforts of old-time trail drivers. Their neckerchiefs cut down the dust they breathed in, but couldn't keep their clothes from filling with grit, which constantly worked like sandpaper against their skin. Judging from today's conditions, those trail hands were a long way from comfortable.

After branding, we came back to headquarters, cleaned up a bit, then had ourselves a barn dance. Sam had planned a barbecue outside, but the wind drove us inside the big metal barn where tractors and other equipment are stored. I say "dance," but it was only Sam doing the two-step with Kathy or Abby, while Leigh Ann played guitar and sang. Pete and Leigh Ann Marez are Sam and Kathy's best friends here, with children about the same age. Pete is an animal control specialist—coyote hunter—for the state. "An assassin," teases Sam. Leigh Ann is an assistant brand inspector, another state appointment. In addition to their state jobs—"Nine to five, home by noon," quips Sam, no fan of government waste—Pete and Leigh Ann train horses. Pete ropes well, and so does Leigh Ann, who often competes with Sam in team roping events. She's also quite accomplished on guitar and vocals, looking every bit the country star with her jeans and boots and honey-colored hair. Souli shook his head and said, "She's pretty, she can sing, she can rope . . . and she married Pete!" Kidding, of course—a backhanded compliment on Pete's good fortune. Leigh Ann seems to be a Renaissance western woman, one who demonstrates traditional feminine virtues at the same time she can ride and rope better than most men. Our own Annie Oakley or Dale Evans. I reminded Souli of the old cowboy wisdom, "Some men's wives are angels . . . others are still alive." Yet there's no denying it, Leigh Ann sang like an angel and made our little celebration sweet.

While she played and sang, Pete fried a mess of fish he'd caught in Santa Rosa Lake. He'd made himself a deep-fry rig, using an old disk blade that he heated with propane—a kind of cowboy wok. Along with the fish, Pete fried up some Rocky Mountain oysters, fresh from the Cowden Ranch branding pens. All these ranch women—Kathy, Leigh Ann, and Jean—seem to be particularly fond of calf fries, or at least claim to be, smiling like angels.

After the feast, Ty and Mark loaded up their horses and left for home. They're both good cowboys, and Ty seems particularly committed to making himself into an accomplished cattleman. He's more ambitious, industrious, and curious than your average cowboy. The distinction is always clear: cowboys work for cattlemen, cattlemen work for themselves. Or, as my grandfather said, for the bank. Both cowboys and cattlemen generally subscribe to a common code of behavior, and ranching doesn't split neatly into labor and management, but they are distinct classes. The rancher, the cattleman, owns property—land and cattle. Given that, he looks for proven methods and values from ranching history and for new techniques and products that will help him succeed; his scope is larger. Since before the founding of the JAL Ranch a century ago, back to the days of the "cattle kingdom" after the Civil War, Cowdens have been cattlemen. Although Sam worked today in the wind and dirt as hard as any hired hand—he looked like the sorriest man in the outfit when we were done—his character and command are always evident.

May 12, 2000—Gathered and sorted Conchas today: Sam, Herman, Juan, Souli, Jerry, me, and Leonard Lujan, who returned to help and visit with old friends. I started from the north corner of the pasture, where I chased off a couple of coyotes and pushed a solitary bull toward the rendezvous at Big Red. Junipers are scattered across the upper part of Conchas, giving way to a draw chock-full of cholla, the cactus so thick in places it's hard to pick your way through. A good pair of chaps would come in handy, but they seem like an affectation for me, since I ride so infrequently.

Herman, Juan, and Leonard started down where the pasture drops off the mesa toward Alamito, while Sam, Souli, and Jerry took the west side. We all arrived at Big Red about the same time, a good gather. The sort went smoothly as well, with lots of dries and heavies; we cut out 53 cows and left 124 pairs. Sam is troubled that so many cows are either open or late—30 percent of that bunch—but can't quite figure out the cause or the solution. All his cows were pregnancy-tested in the fall, so Sam thinks the dry cows aborted calves for some reason. He tries to give his cows adequate nutrition and has added vaccinations for leptospirosis and vibriosis—diseases relatively new to this territory—but 6 to 7 percent of his cows fail to produce live calves. Possibly related to that are the number of cows that calve late, suggesting that they have trouble conceiving. "My schedule's all screwed up," Sam says. "It may take me years to get these cows calving together early in the season. You can delay them easy enough, but it's hard to move them up. You just about have to hold them over a whole year, and who can afford to do that?" Increasing and timing his calf crop are a couple of those long-term management issues Sam wrestles with while he's addressing immediate day-to-day problems: a cow with cancer eye, a water tank going dry, a pickup with a bad bearing. All of it is his responsibility.

Lucky for Sam he has Herman, who is the best hand one might imagine and knows the land, cattle, and operation inside out. There's little he can't do. He's excellent with the machinery—road grader, pickups, pumps—

and works hard and happily. While you don't see much of Debbie around headquarters, Herman is everywhere—out at the barn with the horses, loading feed into one of the pickups, working on equipment at the Bullpen. The ranch without Herman is practically unthinkable.

For a long time it was the same with Leonard Lujan, who lived with his family over at Alamito, working seven years for Sam's father, then twenty for Sam. "Leonard is old-school," says Sam. "Daddy was always 'Mr. Cowden,' though he calls me 'Sam.' Not once was he ever late; not once did he leave early. Always worked hard, always got it done." Of the many remarkable things about Leonard is that he did all his ranch work—feeding, riding, fencing, fixing windmills, whatever—with a wooden leg. Like Chester on *Gunsmoke*, Sam stumps around the bunkhouse imitating Leonard, humor that acknowledges yet does not belittle Leonard's condition. In the same vein, Sam and Herman demonstrate how in later years Leonard shook so badly he could hardly wield a branding iron or how he'd pull on his ear as he talked with you, so that his left earlobe hung noticeably lower than the other—all the tics of a man they had spent decades with, week in and week out. I felt lucky to spend a few days with him during branding.

"My grandfathers on both sides were Apaches," says Leonard, "who married Spanish women." Like many New Mexicans, Leonard carefully and proudly distinguishes his ancestry as Spanish, not Mexican. Leonard himself doesn't have classic, rather angular Spanish features—he's square and blocky from his Indian blood. And he's proud to be Native, especially Apache. "My paternal grandfather didn't know when or where he was born," Leonard says, "since his parents were killed in the battle of Cañoncito, there at Apache Canyon. They were part of Geronimo's band." I wonder if that's true or the sort of family myth attractive in the Southwest—Anglos with Texas Ranger ancestors, Spanish descendants of conquistadors. This was Apache country, before Comanches drove the Apaches south and west; Mescalero Apaches have a reservation down south in the Guadalupe Mountains, and the Jicarilla Apache reservation is in northwestern New Mexico. Geronimo was a Chiricahua Apache, and most of his activity was in Arizona or Mexico. I don't doubt Leonard's Indian ancestry, just his historical accuracy.

Along with his Spanish and Apache background, Leonard also has Anglo connections; he fought for the United States in the Korean War, where he lost his right leg. "I don't have no regrets," he says. "My feeling is, I would have lost the leg anyway. The only thing that troubles me is leaving the folks who were taking care of me when I enlisted. They didn't want me to go, and after I got back, we were never close again." He would have lost the leg anyway?! Leonard's fatalism and stoicism remind me of the books I've taught in Native American lit classes—N. Scott Momaday's *House Made of Dawn* and Leslie Silko's *Ceremony*—which both have Indian veterans as protagonists, men far more scarred than Leonard Lujan with his missing leg. Somehow he's able to accommodate all three cultures that have intersected and often clashed throughout New Mexico history—Spanish, Anglo, Indian. "Tell me, Leonard," I ask him. "Are you an Indian cowboy or a cowboy Indian?" He smiles and doesn't answer. He doesn't have to, this man whose life has been so complex yet so humble: Apache-Spanish veteran of the Korean War, cowboy with a wooden leg on a ranch in the heart of New Mexico. Did I say that he lost a child, a daughter killed in a rodeo accident? Marla Lujan Memorial Arena in Santa Rosa is named in her memory. Sam's father was also a vet—Guy Tom Cowden enlisted at seventeen and served in the Pacific in World War II—so the parallels are close: ranching, military service, lost child. No wonder Leonard also called this ranch home for so long.

Of course Leonard has the longest cultural ancestry here, and deep ties to the land—deeper than those of Herman with his centuries-old Spanish background, deeper than those of Sam, who holds title on the ranch. This afternoon we went to see some "Indian writings" that Leonard's son had discovered while riding one day. Sam, Leonard, and I drove to the site. On a large rock face just below the mesa rim, impossible to see from above, are a series of petroglyphs etched into the rock. I could make out some figures easily—deer, bird, corn, rain clouds—but the whole story, if there is one, escapes me. I plan to come back and photograph the figures and try to determine who might have inscribed them, and when, and what they might mean. It's likely there's no coherent narrative but rather a suggestion of events, an accumulated story. It was thrilling to see them and to imagine the "primitive" lives led there on the mesa, looking west toward the mountains while tribal members chipped arrowheads and guarded a small stand of corn. Those past, nearly lost lives still resonate today, however insignificant they may seem to American culture as jets streak overhead to and from Albuquerque and other cities far from these high plains and isolated mesas. I can't help but feel that people living in cities, doing important and valued work, are nevertheless the ones out of it. Here on the ranch, where constant, concentrated human presence and endeavor have not overwhelmed the physical past, where Indian petroglyphs and Spanish rock-house ruins still survive, I can see the wholeness of time and place. Sure, there are historical sites in Boston and ancient graveyards in New England, but relics here stand in greater relief, on their own in their own resting places, not removed and preserved, however thoughtfully, in museums. I stand where he stood who carved the rain cloud into the rock centuries ago.

May 13, 2000—We finished branding today, working the herd up on the mesa at Three Corners pen, where Conchas comes together with North and South Sabino pastures. It was cool this morning, with a light breeze— ideal conditions. The cows in Conchas have been late, and their calves are fairly small, making it easy on the flankers. And we had plenty of help: Sam and Herman, Souli and Jerry, Leonard and Juan, and three local cowboys. Ernest Copeland is a preacher and a small rancher, Hobby Kaufman used to manage a car dealership but now cowboys, and Kasey White is the fourteen-year-old son of a local rancher. Of the three, Kasey may be the best cowboy, a first-class roper who competes successfully in team roping events. At breakfast he was quiet and modest, where Hobby was affable and Ernest seemed wary. Nobody knows what to make of me on first meeting, a writer here working on a book, a damn professor! I'm a tenderfoot, as far as they know; only my relation to Sam gives me any credence before I have a chance to show myself an able hand. I don't really feel I have anything to prove to anyone except Sam, who has come to realize that I do know cattle and how to work them. I have a world to learn about his operation, about Western cowboys and ranching, the details of life in the saddle out here, considerations of home and range, but I am not entirely ignorant.

The Cowden kids and Souli's son Seth were at the branding, as they had been all week. Sam and Kathy want to involve them in the life of the ranch, and so they helped pen cattle and were assigned manageable branding jobs: Abby gave vaccinations, Hannah ran irons to Herman, Guy helped me earmark or implant calves. With a tool resembling the hole punch that schoolchildren use on their binder papers, I snipped a V-underbit on the right ear of heifer calves. Earmarks have long been used in connection with brands to identify cattle; there are scores of marks, from slickears (uncut) to swallowforks (Vs cut on the tip of the ear) to jinglebobs. The last type of mark was made famous by John Chisum, whose enormous nineteenth-century ranch was along the Pecos near present Roswell; his cattle's ears were slashed and allowed to dangle like pendant earrings. Sam's mark is comparatively simple; neat little bloodless wedges of their ears accumulated in the dust, and the kids took to collecting them, like city kids might collect pop-tops from soda cans. "You could make a necklace!" I kidded Abby and Hannah, who frowned at first, then seemed

Slickear

V-underbit the left

Jinglebob right & left

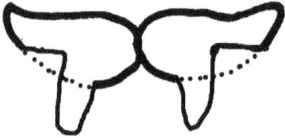

Tip the right, swallowfork the left

Sample earmarks

to consider the possibility. Steers are branded but not marked; they won't remain long enough to need extra identification.

Kathy was typically helpful and efficient: looking after the kids, taking photographs, getting the snack ready when the cowboys took a break halfway through the branding. Sam paces the work well, never getting in a hurry, never trying to do too much, avoiding stress as much as possible. The effect is to make branding an enjoyable event for his family and friends who come to help. All the cowboys work hard and in good humor. Those here to "neighbor" are rewarded most by getting to rope and drag calves, and Sam will reciprocate and help them with their own cattle. If you are not a large, corporate ranch like the Singleton outfit, neighbors are the only way to meet the labor crunch during branding and shipping seasons. Neighboring also provides company and entertainment; rodeos developed out of just such an environment, as working skills became contests between assembled cowboys. Supposedly the first rodeo in Texas was in Pecos on July 4, 1883, when a number of ranch foremen got their outfits together for some sport, shortly before the Cowdens moved the JAL herd to West Texas.

Then there was Hannah's take on branding. "I like it when the cowboys come," she said at the big midday meal that Kathy provided after the branding was done, "because we get good things to eat!" Everyone laughed, then teased Kathy about her everyday cooking, which is plenty good, if not always so replete as a branding dinner. Brisket, green-chile casserole, beans, homemade bread, salad, dessert—no wonder Sam doesn't try to work much after noon.

This evening the bunkhouse crew is collecting their equipment, getting ready to head home tomorrow. Jerry Young and Sam went out to the barn to check the sick calf Jerry retrieved after the first gather; it was unable to keep up and had to be left behind. New calves can suffer from scours, a condition marked by diarrhea and weakness that can render them unable to nurse; without treatment, they often die. For all his tough-guy talk, Jerry has a soft spot, evident in his attention to this calf, which he brought into headquarters and has been doctoring and bottle-feeding two or three times every day. Sam is trying to persuade Jerry to take the calf home with him, but Jerry won't have it. The truth is, neither of them really want the burden of long-term care, but letting it die untreated seems too heartless. It reminds me that my sister Suzanne one summer at the ranch adopted a dogie milk-calf, Calfus, who began following her all over. We've taken to calling this calf "Jerry," as in "Jerry's out with Jerry in the barn." I expect that Jerry will remain behind after Jerry heads back to Colorado.

Souli will load up Seth and his horses and dogs and drive back to Rocksprings early tomorrow morning. He's an interesting character—quick, funny, well-read, complex. Now that he's stopped drinking entirely, Souli's life is less chaotic, but you are always aware of his barely contained energy, both physical and mental. I expect he had a lot in common with my grandfather, when Guy was drinking and carousing along with his Midland ranching buddies. For both, devotion to their families finally overcame self-indulgent and destructive tendencies that threatened to ruin them, and they quit entirely. The image of a wild cowhand riding into Dodge City after a long trail drive is not far off for Guy and Souli; something races in the blood and is restrained with great difficulty. At one point Guy told Sam, "About the drinking, I hardly had a chance," so pervasive was the spirit of the Old Wild West in its inheritors, men forced to settle down on ranches and in town with families. Only a generation before Big Guy, there was frontier territory, Indian wars, cattle drives, the romantic stuff of dreams. It was hard to give that up.

So that's it for branding in year 2000, although I've got lots more to find out about ranch life and practices out here. Tomorrow I'm off to Texas: Lubbock, Midland, and Palo Pinto. I'll be going backwards in time, backtracking the route the family followed to arrive here and now—the Cowden Ranch, New Mexico, at the end of the twentieth century.

Gone to Texas

*He would be a rash prophet who should assert that the expansive character
of American life has now entirely ceased. Movement has been its dominant
fact, and, unless this training has no effect upon a people, the American
energy will continually demand a wider field for its exercise. But never
again will such gifts of free land offer themselves.*

Frederick Jackson Turner,
The Frontier in American History

Why did they head west? people ask, and I suppose that's my first
task—to discover, or rediscover, what drew or pushed the Cowdens
to the frontier. In the first half of the nineteenth century you often saw the
phrase "Gone to Texas," or "GTT," scratched on doors of abandoned
cabins in the Deep South or scribbled on bankers' account books up
North. GTT: someone had skipped out on a failed farm or impossible
debt or trouble with the law; someone had gone to Texas, where the
country was wide open and you could start fresh. In fact, you had to start
fresh. There was no alternative. You had to make something new of
yourself and something new of the territory. All sorts headed west—
adventurers and impresarios, dreamers and farmers, loners and families,
even men "on the outs with the law." You had a fighting chance in Texas.
What sort were those first Cowdens?

GTT! Anything was possible!

The first European to tour the American Southwest, across what would
become Texas and New Mexico, was Álvar Núñez Cabeza de Vaca, the
shipwrecked Spanish adventurer who wandered lost from 1528 to 1536
with three companions. Toward the end of his journey, Cabeza de Vaca

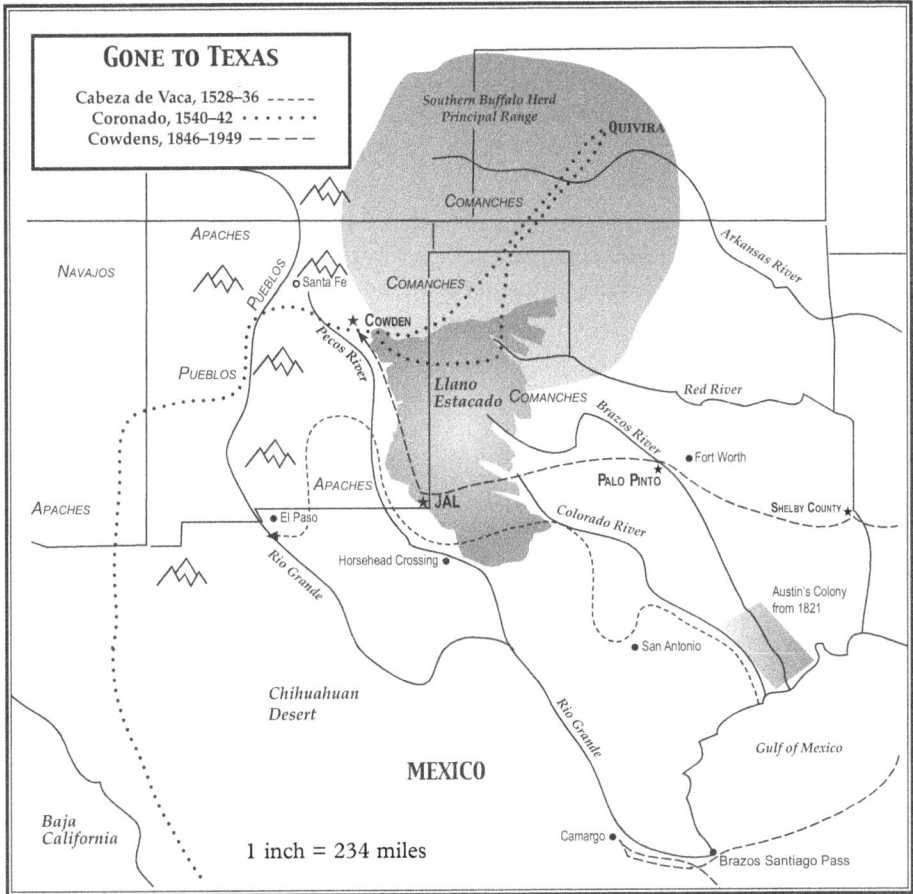

GONE TO TEXAS

Cabeza de Vaca, 1528–36 -----
Coronado, 1540–42 · · · · · ·
Cowdens, 1846–1949 — — —

*Southern Buffalo Herd
Principal Range*

QUIVIRA

COMANCHES

APACHES

NAVAJOS

PUEBLOS

○ Santa Fe   COMANCHES

★ COWDEN

Pecos River

PUEBLOS

*Llano
Estacado* COMANCHES

Brazos River   Red River

● Fort Worth

Arkansas River

APACHES

APACHES

★ JAL

● El Paso

PALO PINTO

SHELBY COUNTY ★

Colorado River

Rio Grande

Horsehead Crossing ●

Austin's Colony
from 1821

*Chihuahuan
Desert*

Rio Grande

● San Antonio

MEXICO

Gulf of Mexico

*Baja
California*

1 inch = 234 miles

Camargo ●

Brazos Santiago Pass

traveled up the Pecos River, through country that Cowden cattle would roam almost four hundred years later. His *La Relación* presents accounts of lands and peoples and experiences that seem the stuff of bad dreams: "We went naked through all this country; not being accustomed to going so, we shed our skins twice a year like snakes. The sun and air raised great, painful sores on our chests and shoulders. . . . The region is so broken and so overgrown that often, when we gathered wood, blood flowed from us in many places where the thorns and shrubs tore our flesh."

The information Cabeza de Vaca carried back to New Spain included facts hard to believe and tales that many wanted to believe. Spanish conquistadores and missionaries—to make reputations and fortunes for themselves, to take territory for Spain, to conquer souls for the Church—mounted expeditions for Cíbola and the other Seven Cities of Gold that supposedly lay to the north. Of these *entradas* the most renowned was led in 1540 by Don Francisco Vázquez de Coronado, at the head of an army of Spanish soldiers, Indian allies, horses, mules, cattle, and sheep. "According to the muster roll . . . ," wrote editors and translators George Hammond and Agapito Rey in *Narratives of the Coronado Expedition,* "Coronado himself was glitteringly arrayed in a suit of gilded armor, with a fine helmet ornamented with plumes. Ten others wore full armor, while about forty had coats of mail. A good many soldiers had various pieces of head armor, and many also were protected by a buckskin coat, the most common armor in the whole force." Four centuries later, Guy Cowden would discover on his ranch an ancient Spanish sword that family legend maintains came from Coronado's expedition. Perhaps. The sword itself—unadorned iron—has disappeared, taken off for historical documentation and lost once again; only its legend remains.

The legend has some basis in fact, although the facts are incomplete and in dispute. Historians know that Coronado made his way north from Mexico to the pueblos on the Rio Grande, then headed east in May 1541 to search for the mythical Indian city Quivira. His exact route isn't known; the village of Puerto de Luna, near Santa Rosa, claims to be the site of a bridge mentioned in Pedro de Castañeda's firsthand account: "The army departed from Cicúye [Pecos Pueblo], traveled in the direction of the plains, which are on the other side of the mountain range. After four days' march they came to a deep river carrying a large volume of water flowing from the direction of Cicúye. The general named it the Cicúye river. They stopped here in order to build a bridge for crossing it. This was completed in four days with all diligence and quickness."

More likely than a crossing at Puerto de Luna is a site at Anton Chico, west of the Cowden Ranch; even more probable is La Junta, where the Gallinas River joins the Pecos, near the southwest corner of the ranch. A bridge below La Junta would have saved Coronado the additional labor of crossing the Gallinas and would have positioned the expedition for the smoothest route to the plains beyond. One can look at the horizon around the Cowden Ranch—mountains and river breaks and plains beyond its borders—to see that the ranch provides a natural path through the country. Observant explorers would take advantage of it; Coronado, despite his failure to find civilizations of gold, was an unqualified success in opening new country, however unpromising it may have seemed. Herbert Bolton wrote in *Coronado, Knight of the Pueblos and Plains:* "Leaving at his back the historic bridge, Coronado and his caravan marched leisurely eastward across a wide plateau broken by boldly scarped mesas sprinkled with scrub juniper, cactus, and other desert plants. . . . The expedition skirted Cerro Cuervo, and probably entered the basin of Pajarito Creek in the vicinity of present Newkirk. Toward the east they could see the imposing line of rampart-like cliffs which gave the vast level expanse ahead of them the name of Llano Estacado (Stockaded or Palisaded Plain)." From the front porch of Cowden headquarters you can see quite clearly, to the southeast, Cerro Cuervo and the imposing "palisades" that Bolton describes; they stand now as then, landmarks of dramatic and severe proportions.

And if Coronado's army did head east across the Cowden Ranch, some heat-struck soldier losing his sword along the way, they may have encountered their first buffalo, the "cattle" of the Llano Estacado that astounded Castañeda: "There were seen also such large numbers of cattle that it now seems incredible . . . so many cattle that those who went in the advance guard found a large number of bulls in front of them. As the animals were running away and jostling against one another, they came to a barranca, and so many cattle fell into it that it was filled and the other cattle crossed over them. The men on horseback who followed them fell on top of the cattle, not knowing what had happened. Three of the horses that fell, disappeared, with their saddles and bridles, among the cattle, and were never recovered. . . . They told the general that in the twenty leagues they had marched they saw nothing but cattle and sky."

Nothing but cattle and sky—besides these the Spaniards discovered only Indians who followed the herds grazing across trackless land. Coronado's swath through the country was, then as now, as much imaginative as physical; the landscape resisted or absorbed most human endeavors, as Castañeda wrote: "Who could believe that although one thousand horses, five hundred of our own cattle, more than five thousand rams and ewes, and more than 1500 persons, including allies and servants, marched over those plains, they left no more traces when they got through than if no one had passed over, so that it became necessary to stack up piles of bones and dung of the cattle at various distances in order that the rear guard might follow the army and not get lost. Although the grass was short, after it was trampled it stood up again as clean and straight as before." But Coronado's expedition did leave traces—bones and legends and perhaps even swords, hidden like the treasure they hunted. Bolton speculated about the horses swallowed by the buffalo stampede: "Perhaps some modern cowboy has picked up and wondered at the metal work of bridles and saddles lost that day, just as other Spanish relics have been found in the vicinity." But more important than whatever artifacts Guy Cowden or others found, was another Spanish legacy: the horse.

After Coronado returned empty-handed to Mexico, the region was left largely untouched, aside from later Spanish settlements along the Rio Grande, San Antonio de Bexar far to the southeast, and missions in distant California. The southern Great Plains remained unsettled land, Indian land: "Over these plains there roam natives following the cattle, hunting and dressing skins. . . . They are by far more numerous than those of the pueblos, better proportioned, greater warriors, and more feared. They go about like nomads with their tents and with packs of dogs harnessed with little pads, pack-saddles, and girths." Castañeda's account held true for centuries after Coronado departed, with one important difference—Spanish exploration brought the horse to North American Indians. Lost and stolen, mainly from Don Juan de Oñate's colony (San Juan de los Caballeros, established in 1598 along the Rio Grande north of present-day Santa Fe), horses changed the lives of Plains tribes and, consequently, the lives of all who encountered them.

Dispersion of the Spanish horse among North American Indians, by way of trade and tribal warfare, intensified after the Pueblo Revolt of 1680 drove the Spanish from the upper Rio Grande valley, their stocks of horses left behind. Walter Prescott Webb wrote in *The Great Plains:* "Kiowa and Missouri Indians were mounted by 1682; the Pawnee, by 1700; the Comanche, by 1714. . . . How much earlier these Indians rode horses we do not know; but we can say that the dispersion of horses which began in 1541 was completed over the Plains area by 1784." This hardly captures the wonder and transformation among Plains tribes, for whom horses became indispensable, changing the quality of their life, if not its fundamental nature. As Webb noted: "The horse did not introduce new qualities into Plains culture. The true Plains Indians already possessed the traits that they continued to exhibit. They were nomadic, nonagricultural, warlike people who depended primarily on the buffalo herds for sustenance. What the horse did was to intensify these traits. The Indians of the Plains became more nomadic (that is, they ranged farther), less inclined to agriculture, more warlike, and far better buffalo hunters than they had been before. . . . The horse ushered in the golden era of the Plains Indians."

In this golden era—and how ironic that conquistadores did not gather, but instead dispersed, "gold"—certain tribes flourished as hunters and warriors and raiders on horseback. Tribes and peoples who mastered the horse mastered the plains. Foremost among these were the Comanches, whose movement into the southern Great Plains was concurrent with the dispersion of the horse; over time Comanches and mustangs intertwined to become one force of nature. Originally part of the Shoshone tribe of the northern Rocky Mountains, Comanches broke away and drifted south and out onto the plains, "in part forced by the incursion of more powerful tribes equipped with muskets coming from the northeast," as Ernest Wallace and E. Adamson Hoebel pointed out in *The Comanches: Lords of the South Plains*. "However," the authors continued, "the southward migration cannot be explained altogether as the result of superior enemy pressure alone. The Comanches had obtained the horse, being among the first tribes of the plains to do so, and the desire for a more abundant supply of horses was certainly an important motive for moving closer to the source of supply: the Southwest. . . . Furthermore, once they had horses, they could really launch forth on a buffalo-based economy, and the buffalo was ever present to the south." Comanches moving down into Texas and New Mexico in turn forced Apaches in the region farther south and west; together these two fierce tribes discouraged settlement by the Spanish and others for almost three hundred years. Comanche and Apache influence sat over the region like a huge ridge of high pressure that keeps rain away; Texas and New Mexico lay barren of settlers, as if waiting some great storm.

The interior of Texas and New Mexico was too distant and too perilous for colonists, friars, and soldiers from New Spain—looking out from a mission or presidio, they saw the open landscape of mesquite and chaparral as the source of little but terror. They kept to the Rio Grande, their haciendas mostly south of the river. Over decades and generations, mounted Indians took full possession of seas of grass and buffalo, the plains theirs by virtue of continual, if nomadic, habitation. Nevertheless, forces were at work that would open the region to others. After the American Revolution, the pace of westward Anglo-American exploration and settlement accelerated: the Louisiana Purchase in 1803 was followed immediately by Lewis and Clark's expedition and by Zebulon Pike's journey to the Southwest, including New Mexico. Around the margins of the "empty" southern plains was restless, isolated activity—mountain men and traders and explorers and settlers. In 1821 three dramatic changes took place: American traders opened the Santa Fe Trail from Missouri, Mexico won its independence from Spain, and Stephen F. Austin's colony began drawing Anglo-Americans to east Texas.

For Austin and others from the increasingly crowded United States, Texas was a world of opportunity—fertile, open, virtually free. Public land that cost at least $1.25 an acre in the American South cost just one-tenth of that in Austin's colony. And if you claimed you were going to raise cattle as well as crops, you could get a square league—4,428 acres. "Gone to

Texas!" were the last words of enterprising or desperate men back in Missouri and Virginia, Carolina and Georgia, Tennessee and Kentucky. "Backwoodsmen—a bold and hardy race, but likely to prove bad subjects," wrote the British minister to Mexico in 1825, six months before the birth in Georgia of William Hamby Cowden, who would eventually join the Texas migration. "Six hundred North American families are already established in Texas; their numbers are increasing daily, and though they nominally recognize the authority of the Mexican government, a very little time will enable them to set at defiance any attempt to enforce it." *Empresario* Austin himself acknowledged he had "a set of North American frontier republicans to control, who felt that they were sovereigns, for they were beyond the arm of the government, or of the law, unless it pleased them to be controlled." By 1830 the twenty thousand Anglos outnumbered Hispanics five to one, and the Mexican government sought to rein in defiant Texans, who had spit the bit and were running away with the colony. What followed inevitably was revolution and independence in 1836, which captured the imaginations not only of Texans but of countless others back in the United States.

Although the new Republic of Texas claimed the territory north and east of the Rio Grande—including half of New Mexico and parts of Colorado and Oklahoma—its fertile coastal lands attracted most new American settlers. They came pouring in from the border states and the Deep South, as T. R. Fehrenbach noted in *Lone Star:* "Ninety percent of the immigration into Texas was composed of native-born people from the old South. . . . Texas was basically Southern in its cultural patterns; it was entirely agricultural, and it was a slave state. But it had one major difference from the southern states: Texas still possessed a long and savage internal Indian frontier." However attractive Texas appeared to land-hungry folk back in the states, it remained in conflict, its Anglo immigrants scrapping with both Indians controlling the western plains and Mexicans south of the Rio Grande. And once Texas was annexed to the United States in 1845, full-scale war broke out, drawing more Americans to the new land— the Mexican-American War would bring the first Cowdens to Texas.

William Hamby Cowden, a month past his twentieth birthday when the United States declared war, needed no prodding to join up. Years before, on the family farm in Newton County, Georgia, he heard the news of fallen and triumphant heroes at the Alamo and Goliad and San Jacinto, battlefield accounts that fired a boy's imagination. Later came word of Texas Rangers fighting Indian wars, events far more exciting than chores—hoeing fields or tending cattle—in a land already legendary. As soon as they heard that fifty thousand volunteers were needed to fight, Bill and his younger brother George Franklin Cowden enlisted. Volunteer companies were being raised in Georgia, but perhaps to improve their chances, the two boys traveled west to Alabama, where they signed up in Company I, First Alabama Volunteers. They were going to fight for their country, to see the world. They were going to Texas.

Like most early volunteers, the Cowden brothers enlisted for one year, drawing the same pay as regular soldiers but furnishing their own uniforms. They did not receive, as did later volunteers, enlistment bonuses of free public land. After some training in Alabama, where the volunteers elected their own officers—including Bill as lieutenant and young George as captain—the Alabama regiment sailed from Mobile on the steamboat *Fashion*. The Cowden boys had never seen the sea, much less passed time aboard a ship wallowing through a Gulf of Mexico storm. A diary entry by another young volunteer reveals the rite of passage: "Last night was rather a mixture of odd scenes. The excessive tossing of the vessel affected most of the men with seasickness and it was one continual sound of pukeing, spitting, groaning and laughing mixed with the tumult of the gales. . . . Some of the companies had not secured their provisions with lashings to the vessel and during the night it caused quite an uproar of barrels sliding and dashing across the deck and Uncle Sam's provisions were scattered in glorious confusion . . . by a heavy sea which struck the old ship broadside and laid her on her beam ends. At the same time the anchor chain was dashed through the fore hatch. For a short time, until the ship righted, the noise of box, chain, barrels and all things movable on the ship was terrible and every man was hanging on to his berth to prevent having his brains dashed out."

On July 4, 1846, their ship arrived at Brazos Santiago Pass, near the mouth of the Rio Grande, where William Hamby and George Franklin Cowden first set foot on Texas soil. Their camp was on Brazos Island, which Justin Smith reported in *The War with Mexico* "consisted of low hills on the side toward the mainland, a swamp in the centre, a wreck-strewn beach on the outer side." The regiment moved inland to a camp overlooking the Rio Grande, "a spot fit only for the snakes, tarantulas, centipedes, fleas, scorpions and ants that infested it." By late August they were transported upriver to Camargo, Mexico, where "every breath of air raised a stifling cloud of dust from the dried and pulverized mud." The report continued: "Barren hills of limestone cut off the breeze to a great extent and concentrated the fierce heat, frequently sending the mercury in 'this hottest of all hot places,' as a soldier called the town, to 112 degrees. Scorpions, tarantulas, mosquitos and centipedes abounded. There was a plague of small frogs." The First Alabama languished in Camargo until nearly Christmas, increasingly sick of life in camp: "They had come for glory and a good time, and were having neither. They wanted to do something, and to do it at once or go home. One at least of them believed that assignment to garrison duty would have led to general desertion. Wherever they were, they wanted to be somewhere else. Having come to gamble, with their lives for a stake, they clamored to have the game begin."

Sent next to garrison Tampico, the impatient soldiers at last joined General Winfield Scott's force, invading Veracruz on March 9, 1847. After nine months of waiting, the Cowden brothers went into battle. The siege of Veracruz did not have the ring of glory, however, as General Scott himself noted: "Although I know our countrymen will hardly acknowledge a victory unaccompanied by a long butcher's bill I am strongly inclined—policy concurring with humanity—to 'forego their loud applause and aves vehement,' and take the city with the least possible loss of life." The butcher's bill was short—only nineteen Americans died in the siege. Soldiers suffered more from the elements and disease than from Mexican fire; the greatest threat at this point was *el vómito*, yellow fever, due along the Mexican coast at any time. Shortly after Veracruz surrendered, the Alabama regiment marched off to capture Alvarado, where William Cowden turned twenty-one years old.

Scott's army fought its way off the feverish coast, up into the Mexican highlands, routing Santa Anna's troops at Cerro Gordo, though the First Alabama missed the pitched battle by one day. Not long afterwards, nearing the end of their twelve-month term and unwilling to reenlist, the Alabama and other volunteer regiments were sent home early to avoid high yellow fever season at Veracruz. Justin Smith wrote of their departure: "The soldiers had learned what campaigning really meant. They had been allowed to go unpaid and unprovided for. They had met with hardships and privations not counted upon at the time of enlistment. Disease, battle, death, fearful toil and frightful marches had been found realities. . . . They 'sighed heavily' for home, family and friends." On May 6, 1847, the volunteer regiments headed back to Veracruz, boarded a ship for New Orleans, and by the end of the month Lieutenant William Hamby Cowden and Captain George Franklin Cowden had arrived home.

The two young brothers' military service was brief, but the Mexican War seasoned them in ways that would make possible, even probable, a life on the frontier of Texas. They saw both heavy fighting and Mexican *guerrilla* ("little war" in Spanish) tactics; in battle and camp they faced acute and chronic conditions unthinkable back in "civilized" Georgia or Alabama. When they came through, they were tough and confident, veteran army officers. Yet, as volunteer rather than career soldiers, they maintained their connection to civilian life, returning to the pull of home and family, like Cowdens to follow. Life in dismal army camps was not in their future. Still, the risk and adventure had been considerable—that leap disposed and prepared them for later leaps into perilous territory.

And the men from Texas to whom the Cowdens were exposed—rugged, half-wild Texas Ranger companies serving under Zachary Taylor—made a great impression. Riding by the volunteer infantry on horseback, Texans were a breed above and apart: "Hays' regiment of Rangers were irregulars, hardly proper soldiers, who fought like devils, but behaved like wild-men. . . . Men wore every kid of coat—blue, black, long-tailed, and short. They wore leather caps, stained fedoras, and soiled panama hats. Most were bearded. They rode every color of horse. The most excitement was caused by their armament. Each man carried a rifle, a knife, and a rope; some had single-shot horse pistols—but all had two Colt six-shooters prominently displayed in his belt." Fierce, fearless men who had whipped Santa Anna, whipped the Indians, they could do anything. Who would not aspire to their condition? Though rarely boastful themselves, Texas Rangers inspired stories that circulated through idle camps along the Rio Grande, embellished in repeated tellings until they became frontier legends.

Of course legends grew in such a place. The Cowden brothers had heard of Texas, and now they saw it, if only along the Rio Grande. The country before their eyes was ripe for the picking—wasn't that why Americans were there, to seize and hold that land? The Mexican population along the border, in the habit of the times, was dismissed as inept or irrelevant; the Indians of interior Texas were likewise considered inferior, even subhuman. In Anglo eyes the country was unpopulated, available for men of ambition. And for each mile beyond the miserable, regimented army camps, the country assumed greater proportions and virtues; volunteers thinking about eventual return to civilian life had before them an actual dream state—Texas. Around campfires the Cowdens heard reports of rich farm-land along coastal river bottoms, open prairies inland, land given free to white families. And there they were at its threshold, fighting for it—it was *theirs,* by God! Imagine how compelling Texas became. No wonder that the Cowdens, once back in settled Alabama, soon returned.

History—including various and varying family accounts I've seen—is selective and imperious, tending to neglect people and events not eminent or notorious or public. Yet most of any life lacks drama, or seems to at the time. William Hamby Cowden stayed in Alabama long enough to grow restless in a town—clerk in a store and other dull work—but also to find a wife and start a family. In 1848, in Bellefonte, Alabama, he married sixteen-year-old Caroline Martha Liddon, one of eleven sisters. Their children came quickly and steadily after: Mary Josephine in 1849, Willie Jane early in 1851. Carrie was expecting again in 1853 when Bill decided it was high time and took the family west. Somehow—by foot or horseback or wagon, sleeping by the roadside or in rude accommodations, with companions or on their own—William Hamby Cowden, his pregnant wife, and their two infant children made their way across Alabama, Mississippi, and Louisiana, a journey of six hundred miles, fording rivers small and great: Black Warrior and Tombigbee, Mississippi and Red, and finally the Sabine. On the way, Bill and Carrie spoke of realities and possibilities—how many miles they might make by evening, the way the baby was kicking, what they might find on the frontier—but left no written record of the journey.

Possibly on the road in Louisiana, but probably in Shelby County, Texas, William Henry Cowden was born on October 6, 1853, their first son. Settling down in the East Texas piney woods, the Cowdens were part of an enormous migration into the new state, described by Rupert Richardson, Ernest Wallace, and Adrian Anderson in *Texas: The Lone Star State:* "Annexation and the continuation of the liberal land policy brought a veritable stream of homeseekers. The lure of cheap land in Texas was second only to that of gold in California. . . . By 1860 the population of the state had jumped to 602,215, almost three times that of 1850. Of the 1860 population, almost three-fourths had been born outside of Texas." Fehrenbach speculated about what drew the settlers: "Immigration into Texas was part of the expansion of the South itself; it was not an expansion out of the adjacent states of Louisiana or Arkansas, but by families who leapfrogged from Alabama or Tennessee. One-half the white population came from these two states. . . . They wanted to get away from the slave plantations, with which they could not compete; they could most easily acquire land on the far edge of settlement, and there was, noticeably, in these people an urge toward the far frontier. They took their cabin lights to the edge of Indian country."

Among those Alabama thousands was Bill Cowden and his growing family; his brother George also arrived in Shelby County at some point. There the two brothers demonstrated, as they had in the Mexican War, their capacity for adventure and risk and effort, raising crops—corn to consume, perhaps a little cotton for cash—and cattle in country that was new, but not foreign, as cultural geographer Terry Jordan remarked in *Trails to Texas:* "The Texas Piney Woods region, while somewhat inferior to the Coastal and Blackland prairies as a cattle range, was almost exactly what the pine barrens herders from the Lower South had known back east."

Well watered, with stands of timber and occasional open grasslands, Shelby County may have seemed too much like the Georgia and Alabama that the Cowdens left behind. And too crowded as time passed: new settlers kept arriving, and in December 1855 Carrie gave birth to another son, George Edgar Cowden. Along with the growing family was a growing herd of cattle—thirty head when East Texas evidently could contain them no longer, and once again the Cowdens loaded their household possessions on wagons, piled in the children, and took off, moving west. Hunger for land, or perhaps for the very frontier itself, drew them deeper into Texas, over two hundred rough miles up the twisting Brazos River to the brand-new county of Palo Pinto.

Tumbleweeds
May 2000

May 14, 2000—I drove off the ranch this morning, through with long days of gathering and branding, eager to see new country. I followed the Pecos down to Fort Sumner, where Navajo and Apache Indians were once held prisoner on the Bosque Redondo Reservation. From there I went east across the Staked Plains, an old Comanche route now a highway passing through Clovis, across the state line at Texico, then through Lariat, Muleshoe, and Littlefield, proud birthplace of country singer Waylon Jennings. After Littlefield came Roundup and Lubbock, proud birthplace of singer Buddy Holly. Lubbock is the birthplace also of writer Michael Pettit. My parents were living in Midland, but the hospital there was closed for renovation, so my infant footprints were inked onto a birth certificate from Lubbock General Hospital.

I drove in past farm supply and oil-field service shops, auto parts stores, barbecue joints, pawnshops, and more Bible supply houses than I could count. I took a turn down Buddy Holly Boulevard, wondering what led Holly, Waylon Jennings, and Roy Orbison down in Wink, Texas, to become musicians. Despite singing cowboys like Gene Autry and Roy Rogers and the Sons of the Pioneers, despite the whistling wind, the plains do not seem musical. It seems all song would be hopelessly lost in those wide-open spaces. Perhaps that's why song arises there, as it does deep in the Appalachian Mountains, a cry against an isolated, lonesome landscape. Even the most manly of souls—a cowboy—might have reason to sing, as he rides night herd under the vast starlit sky, soothing spooky cattle with some soft lullaby.

Lubbock arose near the site of old George Causey's pioneering camp in Yellow House Draw, as I found out this afternoon at the library. Causey was a buffalo hunter who moved with his brothers from Kansas, hunting around the Panhandle before arriving in 1877 at Casas Amarillas—the Spanish name for a yellow, flat-topped bluff pocked with caves—where they built a sod home near a reliable spring. From there they ventured out with their Sharps .50 rifles to hunt buffalo, helping to complete the slaughter of the southern herd. In 1877 his crew sold eleven thousand hides; Causey alone is reported to have killed forty thousand buffalo in thirteen years of hunting. By 1882, buffalo hunters had efficiently put themselves out of business on the southern plains, at the same time destroying the basis for Plains Indian culture. Causey and others thus opened the gates for the cattlemen of Texas, who began moving west with their herds into the new void.

Yellow House Draw is a headwater tributary of the Double Mountain Fork of the Brazos River, three hundred miles upstream from where the Cowdens in Palo Pinto County raised cattle. The Cowdens, looking for greener, or at least more open, pastures, eventually moved their herd from the main drainage of the Brazos up the Double Mountain Fork, at about the time Causey abandoned buffalo hunting for ranching. Causey captured and raised mustangs at first, turned to cattle, and soon began digging wells and rigging windmills on the Llano Estacado. He was an important, if isolate, figure in the eventual development of cattle ranching in West Texas and New Mexico—one of those restless men who always sought the frontier, lured by its possibilities, undaunted by its risks. Causey's late life was hard, after he took a bad fall with a horse and severely damaged his spine, an injury that impaired him for the next three years, until he finally shot himself. The motivation for that act was part physical, part spiritual, as Gil Hinshaw noted in *Lea, New Mexico's Last Frontier:* "Everywhere civilization, which he seems to have stayed one step ahead of, was pushing in upon him. To the north on the Llano, the Four Lakes ranch lay from horizon to horizon; the Mallet Ranch, later the High Lonesome, sprawled to the east of him; immediately to the east and south was the Hat Ranch; and to the southwest and far south the San Simon and JAL ranches occupied the remainder of the Llano and sand hills within New Mexico territory. Causey had no place left to be free."

Lubbock's population is now almost 200,000, and we may have difficulty imagining this area as an empty, hostile environment, but so it was. We may have difficulty imagining a few isolated ranches as civilization, but in important ways they were, a link in the chain of Indian, hunter, rancher, farmer, merchant. George Causey then, now Lubbock. And I am off to J&M Bar-B-Q get some tasty mesquite-smoked brisket and ribs, before returning to my air-conditioned motel room to watch cable TV, effectively nowhere—not Causey's buffalo camp, not Buddy Holly's home-town, not even my own place of birth.

May 15, 2000—Day two at the Southwest Center at Texas Tech University. Appropriately enough for research about the open range, I am the sole patron in the reading room here. Back in Massachusetts, I also had the country to myself—most books I checked out of the library at Amherst College had sat untouched in the stacks for decades. I can look at that two ways: no one's interested, or the word is simply not yet out.

Here's one discovery: the old song "Home on the Range" was a literal appeal, popular among buffalo hunters who had slaughtered out the herds in Kansas:

Oh, give me a home where the buffalo roam,
Where the deer and the antelope play,
Where seldom is heard a discouraging word,
And the skies are not cloudy all day.

So said May Price Mosley in *"Little Texas" Beginnings in Southeastern New Mexico,* one of my finds in the library. Many sources I uncover are small, "insignificant" books like county histories, recorded by individuals far from the centers of publishing but exactly at the heart of their own worlds. Although the Cowdens ran a storied ranch, they've remained largely invisible, their lives private rather than public. It's like an associate said of George Causey: "Causey killed more buffalo in one winter on the Yellow Houses than Buffalo Bill Cody killed in his entire lifetime, but Causey didn't have Ned Buntline for a publicity agent." Buntline was a dime novelist and promoter who popularized young Cody and rode to western pulp-fiction fame with his hero. Spin then was as important as spin now. Charles Goodnight first had Texas historian J. Evetts Haley sing his tune; then Larry McMurtry popularized his partnership with Oliver Loving in *Lonesome Dove.* I am here, it seems, to spin the Cowden story.

When I first began research, I sent out a query over the Internet to libraries and bookstores; the response was generous. One person suggested I start with Ramon Adams's *Rampaging Herd,* a definitive bibliography of 2,651 books and pamphlets of the cattle industry. "That will give you some idea of the territory," he wrote, "and it's only the beginning." Someone else recommended the Philip Ashton Rollins Collection at Princeton, which had its start when Rollins, a Princeton graduate with open-range ranching experience, walked into a New York bookstore and said he wanted "every damn book that says cow in it." I quickly realized I had to move in two directions at once: outward after general accounts of western ranching, and inward after the Cowdens' particular experience, past and present. What to include? What to exclude? I was suddenly faced with a world whose breadth and depth were staggering, arrayed before me in the actual landscape and in countless stacks of libraries. Outside and inside were millions of acres, thousands on thousands of days and nights, millions of words.

Here on the plains I've uncovered details about Palo Pinto, where the Cowdens ranched for almost thirty years before they came to West Texas. I need time to sort all the facts and claims, but it's certain Palo Pinto was the source of many cattlemen famous in Texas and southwestern history— Goodnight and Loving, the Cowdens, the Slaughters, and others. It appears to have been a jumping-off point for the eventual movement of ranching across the southern Great Plains. I've got volumes to digest— Rupert Norval Richardson's *Comanche Barrier to South Plains Settlement,* Mary Whatley Clarke's *Palo Pinto Story,* and the Palo Pinto County *Star*'s Centennial Celebration souvenir edition, May 10, 1957. In the *Star* are accounts of early settlers like the Cowdens, descriptions of their marks and brands, evidence that they lived through the major events in the rise of the Cattle Kingdom in Texas. I'm eager to see the country itself, what may have attracted the Cowdens to Palo Pinto in the first place and what may have led them to leave, to head farther west.

The Southwest Center also has an extensive archive of tape recordings of ranching history; I could stay here a week just listening to old-timers describe their lives on West Texas ranches—distances and weather, privations and progress, all the details that compose a life. An interview with Midland native Robert Rankin reveals the following about home on the range: "We lived off game—antelope, prairie chickens, blue quail. . . . Every ranch we'd come to, it was customary to drop [in] off the road, and leave one antelope for them. Every ranch house had a screened-in porch, and that helps suck dry the humidity. You'd hang a carcass of an antelope up, or a milk-pen calf you killed, and they'd just kind of sear over dark. We didn't have as many germs, meat didn't spoil so bad, so quick. We'd just keep it cool out there on that screened-in porch, we didn't have any refrigerator. We had the windmill close to the house, and from the windmill going to this dirt reservoir would be a long, covered trough, with a V-shaped top, with cool running water. You could put your milk in there in buckets and put a brick on them, and in that cool running water in the shade from the sun, they kept fairly cool. That was our refrigeration."

Such details are intriguing for the same reason that *The Swiss Family Robinson* is appealing, with its self-sufficient and inventive responses to a new world. An island in the sea, a ranch on the plains. Exploration and discovery—the West has long been our actual and imaginative frontier, which the Cowdens pursued generation after generation. And now I track them, looking for a footprint, some trace of their passage.

May 16, 2000—I'm in Midland for the first time since I drove away in my grandmother's old Cadillac after she and Guy died ten years ago. The city's changed, creeping north toward Midland College and the new loop, away from the T&P Railroad tracks and downtown buildings half empty because of the oil bust. Midland's got money and activity still, but nothing like the boom fifty years ago when my father came here as a petroleum geologist. At that time the Permian Basin oil patch included George Bush Sr. and others who would make fortunes. Little George, running hard now for president, used to play with my older sister, I am told. But it's hard to sort out the stories, the continuum of fact and legend and myth. Did Buck York really purchase a car over the phone, charging it to his hotel room bill? Who were those wild men who reportedly swiped the train in Mexico for a joyride?

None of this surfaced while I was a child, visiting the quiet home of my grandparents, first the big old house on Illinois and later the brick "ranch style" in the northern suburbs. Guy and Annie Mae had a cook, dinner at noon, a nap afterwards, and occasional spins out to the ranch south of town. They ate a small supper and watched the light fade from a cloudless sky. They slept, woke before dawn, sat in armchairs drinking coffee until breakfast, took a morning trip down to the First National Bank. Their life in town was orderly and calm; only the ranches hinted at some special existence. It was exotic enough to travel from New Orleans to West Texas, but Midland itself had a settled quality, which was the point from the very first.

After it arose on the new Texas and Pacific Railway from a station halfway between Fort Worth and El Paso, Midland became a supply and cultural center for the ranching industry of West Texas and southeastern New Mexico. Down the line, the town of Pecos served the area as well, but never to the same degree as Midland. Ranchers got their necessities from town—windmills, barbed wire for fences, saddles, foodstuffs, clothing, lumber—and drove cattle to the railroad there, shipping them to market in Fort Worth or Kansas. It was, however, a long and difficult haul across the sandhills and plains to and from Midland—seventy-five miles from JAL Ranch headquarters in southeastern New Mexico. Home on the range was a wagon, tent, or dugout. As the Cowden families grew, many of the women and children moved into Midland for schools and churches and other civilized pursuits, according to Lyda Watson, who arrived in 1900: "In those days girls would marry these cowboys and go out on the ranches and live until the children were old enough to go to school. Then they migrated to town."

In town they smoothed out edges rough from their life on ranches. But sometimes the burrs remained. Cowboys working away from town for weeks or months at a time would ride in with attitude. Although Midland was nowhere near as rowdy and dangerous as cow towns at the ends of cattle trails—places like Dodge City and Abilene, Kansas—the spirit was occasionally similar. Men working hard, isolated, and dry relaxed in ways unusual to new arrivals from back East, like Miss Watson. Her brother went out one evening, and when "I realized I was here alone out on these western plains by myself, I was petrified," she said. "I locked the doors and stopped up the keyholes, and the first thing, I heard an opera troupe of coyotes out in the backyard, just a-howling as loud as they could. . . . I had calmed myself somewhat; then immediately after that I heard a cowboy pass by the house, and he was either hilarious or drunk, I do not know which, and he was riding horseback and yelling just as loud as he could. And when he got in front of the house, he fired off a gun. Well, I knew then I was going to be scalped by the Indians."

By the time she arrived in Midland, however, the Indian threat was just a memory, if still vivid to the Cowdens from their days back in Palo Pinto. Indian and buffalo had been replaced with cowboy and Longhorn, and this was town, with a jail and courthouse, churches, banks, stores, saloons, hotels, boardinghouses, schools. The eldest Cowden brother—"Uncle Billy" to relatives and friends—had sold his interest in the JALs, helped establish the First National Bank of Midland, and become its president in 1894. By the 1900 census he was listing his occupation as "Capitalist"; the other Cowdens still listed "Cattle Raising." The JAL Ranch by then was running forty thousand head. Working headquarters were across the line in New Mexico, but business was conducted in Midland, where most of the Cowdens owned homes. In *Open Range Ranching of the South Plains in the 1890s,* Eugene Price wrote, "Pioneer members of the well-known Cowden cattle clan, Bill, Fred, George, John M. and others, were holding down the extreme southeast corner of New Mexico with their renowned JAL Outfit. All these people I knew in Midland—which was home town and trading point for most everyone up to a hundred miles or more to the north."

The next generation of Cowdens was quickly growing; in September 1900 the first son of Eugene Pelham Cowden and his wife, Tennessee Cornelia Moseley Cowden, was born. They named him Guy, after—surprise!—Tennie's sister Henrietta, who everyone called Aunt Guy. Such nuggets are the stuff that surfaces when you dig into memory and records; never in my life would I have dreamed my grandfather was named for a girl.

Guy Cowden inherited a way of life that was changing, as homesteaders moved out onto the plains to attempt small farms on public land previously controlled by ranchers. Free, open range began to disappear, and by 1912, John M. Cowden bought out the interests of his brothers in the JAL. Three years later the great ranch was finally closed down, although most of the family continued to raise cattle on smaller ranges. I expect that my grandfather's childhood was complex. He belonged to a prominent family and a storied tradition, but neither could prevent progress, if that's what we should call the influx of settlers. I can imagine that Big Guy felt a keen sense of loss as the frontier closed before him.

What most lures me about Palo Pinto—I'm driving there soon—is the nature of the country itself, how it resembles or differs from West Texas. That may hold one clue to the migration the Cowdens made, a very puzzling move when you look out the window here at the hot, dry, wind-swept, flat country. Why on earth?! As Lyda Watson said, "It seemed to me I had never seen so much sun and wind and space in all my life. And cattle, and so few people. . . . When I stepped off the train, I felt like I had reached the place where God quit work. It seemed to me that there was more of nothing here than any place I'd ever been."

May 17, 2000—Midland, the "Tall City" rising above the empty plains, its banks and oil buildings shimmering in the heat, visible for miles across the dry, mesquite-covered country. I remember how it would recede into the distance as we drove twenty miles south to my grandmother's ranch, the Rankin Highway littered with roadkill jackrabbits. Once we arrived at the "Rabbit Ranch," my grandfather took us driving through the brush with a .22 lever-action Winchester rifle. My older sister and I were merciless, trying to outdo each other as we alternated shots at big skinny jackrabbits in the shade of mesquite or cactus. I can still see down the barrel the gold bead of the sight, smell the gunpowder, hear the zing of shots that missed high, provoking a twitch from the long pink and gray ears, but no other movement. When our aim was true, the rabbit would leap into the air, then hit the ground kicking and squealing, as my grandfather drove away, saying, "He'll die." Guy Cowden had been through enough lean times in his life not to waste a second bullet on something so insignificant. "Fifty of them eat as much as one cow," he told us, soothing our consciences and stoking our bloodlust—we were doing something for the ranch, the family. And anyway, weren't they everywhere, apparently as limitless as the horizon? It *was* the Rabbit Ranch.

The Youngblood Ranch, its proper name, was twenty thousand acres of poor land, unless you took into consideration the oil wells here and there, and of course everyone did. Oil companies built roads and paid royalties; it was an old axiom that pumpjacks—revolving slowly, their counterweights rising and falling like huge rocking horses—were modern "cattle feeders." Guy and Mushy ran rangy cows and some sheep on the ranch, but the Midland ranch was more important for its minerals than for anything on the surface. The "real" cattle ranch was where the land and water were better, up in New Mexico.

I realize now what a dramatic transformation Guy Cowden lived through, part of a generation of ranchers who saw the open range close down, cattlemen quitting or moving to smaller ranches and making a tough living until the West Texas oil boom transformed their hardscrabble holdings into valuables. By that time Guy had been through droughts and depressions that tempered any tendency toward gaudy demonstrations of wealth, although plenty of that went on among the ranch-and-oil elite, including relatives. Lyda Watson recalled: "The Cowdens at that time were like a band of Israelites—there was no ending of them—and they invited the Mosely girls and me to go out on the ranches with them, and we spent a week just traveling from one ranch to another. It was a liberal education to me. We would be sitting in one of those ranch houses—of course the ranch houses then were not like they are now—and the women would have on house-dresses and broken out with diamonds like the measles, and the men would have on their work shirts and their spurs and a big sunburst. I had never been accustomed to anything like that. We did enjoy it. I remember that first night I played at the church in Midland, and during the sermon a woman came out—well, she was just broken out with diamonds all over. And the only reason she didn't have on any more diamonds was because she didn't have the fingers to accommodate them. She had a diamond barrette in her hair and a diamond sunburst. . . . I was sitting by the minister's wife and I said, 'Well, who is the queen of Sheba?' And she told me she was the cattle queen of the West."

Guy and Annie Mae Cowden were certainly among the Midland elect, but you couldn't always tell. Mushy drove a new Cadillac and oversaw a dignified house in town, but Guy drove a Ford sedan so stripped down it didn't even feature a cigarette lighter. He kept a box of kitchen matches on the dashboard, striking them across the metal face of the dash to light his cigarettes. Scorched black slashes gave the interior a humble touch, matched by scratches along the exterior where he had driven through thorny mesquite and cactus thickets on the ranch. In town, while his wife tended to the family, Guy would haunt the First National Bank of Midland—started by Cowdens and at that time still directed by them—in a spotty white shirt and a sweat-stained Stetson. To the black carhops of the garage, he was "Mr. Cowden" and always their favorite. If he was a cattle king—his uncle had been enshrined in the Hall of Cattle Kings at the Texas Centennial Exposition—Guy had perspective on his kingdom, the same quality that Sam Cowden shows today on his big ranch.

Tomorrow I'm driving to Jal, New Mexico, stopping on my way at the old Dollarhide Ranch headquarters. Part of the original JAL Range, Dollarhide was my great-grandfather Eugene Pelham Cowden's ranch; the family sold the land years ago but still owns the mineral rights. I'll be able to look over the territory, however; my cousin Jon Means is now leasing the ranch to run some of his cattle. It's come almost full circle.

May 18, 2000—Big day. I'm just back from Jal, where the Cowdens sought, and eventually found, their fortune. Not a pretty picture, although I wasn't expecting pretty. West from Midland the land lies flat and treeless, only knee-high mesquite, creosote weed, and yuccas with tall blooms that some say gave rise to the name "Llano Estacado," or "Staked Plains." The highway passed Gardendale, where a forlorn pecan orchard broke the monotony, then through oil fields and production facilities in East Cowden and North Cowden. I turned north just before Notrees, land featureless but for an occasional windmill or pumpjack; past Coyote Corner, the sandhills began, rising beside the highway and extending off into the distance. This was part of the "Great American Desert" that mapmakers placed on nineteenth-century maps—land apparently worthless in the minds of agrarian America, without the abundant water and timber that marked life from the Atlantic seaboard well past the Mississippi River. Here I was, north of Notrees, Texas—nowhere.

The sandhills are covered with shinnery, knee-high shin oaks that hold the rolling dunes in place, or they would drift before the relentless wind and obliterate any human trace. Signs along the highway warned: Loose Sand. No cars came from the other direction; none were headed my way— I'd hate to drift off the road and get stuck out there. The landscape seemed lunar, though the sea was the most frequent metaphor for early travelers across the plains. For 360 degrees, nothing more substantial than sandy hummocks broke the horizon. The sky was so blue it appeared solid.

Eventually the sand gave way to "hard ground," as locals call it, and I headed west through alternating belts of shin-oak sandhills and mesquite flats. Seventy-five miles from Midland, I arrived at Dollarhide, on the Texas–New Mexico border near where my great-grandfather and his brothers ran the JALs outfit almost 120 years ago. I was still nowhere by most standards, though an endless grid of pumpjacks dotted the landscape on both sides of the highway, served by oil-field roads and power lines to the pumps that would eventually power the electric lines. Not a soul was in sight. Union 76 had posted a sign for its North Dollarhide Field that warned: DANGER. H$_2$S GAS MAY BE PRESENT. What does hydrogen sulfide do to you? Something unpleasant. The land itself looked to be mesquite, weeds, and thin soil. A few of Jon Means's big crossbred cattle grazed out in the brush.

Back from the highway half a mile was ranch headquarters: a set of pens, tin outbuildings, a bunkhouse, two ranch houses. The first, and more modest, house was now used by the ranch foreman; the other, built by my great-uncle and -aunt, seemed to be empty. I walked around under shade trees planted near the houses, trying to pick up whatever mystical vibrations might linger. None were evident. A chuck wagon was parked behind the ranch house, a new one in good shape, someone's homage to trail driving and roundup days. I found a collection of old branding irons in a tin garage, but no JAL brands. In fact, nothing echoed of long family history except the empty landscape itself—the horizon seemed more evocative than any building or relic I might run across. I took a few pictures, the last one of Unocal's Dollarhide Well Unit 55-88-D, with its cheery sign:

DANGER	DANGER
NO SMOKING	POISON GAS
CAUTION	CAUTION
THIS UNIT STARTS AUTOMATICALLY	
HARD HATS REQUIRED	

No wonder no one was around. I lit out myself, since I hoped to locate the site of the old JAL headquarters. I had an approximate location somewhere over the line in New Mexico; if I could find where "Old Adobe" once stood, I could probably depend on some chills even in the fierce heat.

First I drove into Jal, New Mexico, itself, the town named for the Cowden brand. Jal is a small oil-field town now suffering through the doldrums of the industry; El Paso Natural Gas pulled out in 1985 and left residents in the dust. Buildings on the edge of town have the look of imminent abandonment; houses need paint; stores need customers. To my surprise, however, near the center of town beside the high school stands a nice new library. Woolworth Community Library is built of brick, set partially into a hillside, much as Cowdens built their dugout houses here when they first arrived. I wonder whether the echo is intentional or just a contemporary architectural response to the climate, using the earth itself to moderate blazing summers and winter winds. Inside, I found another surprise—on the lobby wall hung four huge black-and-white photographs: cowboys branding on the open range; "Jal Boys Catching Horses"; "Outfit Ready for Work," in which nine riders posed before a chuck wagon; and a fourth photograph. In that last one, stretched out in the shade of the chuck wagon, looking for all the world—or at least everyone in Jal—like Wild West desperados, were sprawled JAL cowboys. One of them was reading a book, with another reading over his shoulder, and a third sleeping beside the two literate cowboys. Hanging on the wooden wheel of the wagon was a side of bacon or beef, which the cook was after with a long knife. In each photograph the landscape in the background was flat and spare, like the wide sky above. I could sense the isolation those men must have felt a century ago—perhaps what drew them to the work.

After some research and making copies from *Then and Now—Lea County Families and History,* I left the library and drove to the town park, where I found an oddity I'd read about. Lea County apparently has no running streams and thus not a single roadway bridge. When Jal built its town park, a small lake was dug in the shape of the JAL brand, the center of the "A" an island served by the only bridge in Lea County. A dubious achievement, born out of deprivation and imagination rather than abundance, but what the hell, here was the Cowden brand, immortalized, sort of. I walked across the short, arched wooden bridge and stood on JAL Island—not a deserted island but a desert island.

After that adventure, I drove north from town, into the empty plains again, where I happened across another startling image. Jal was chock-full of surprises. On a rise west of the highway, like mirages in relief against the hard blue sky, I saw the silhouettes of cowboys herding cattle toward town. At first they looked real, off in the distance. But not one figure budged. They held steady on the horizon, realistic but not real. What in the world? I was dumbstruck, though who would I talk to out there anyway? Two riders led the small herd, with another in the middle and a final cowboy riding drag. In the habit of cowmen everywhere, I counted off the cattle: seven cows, six calves. A modest gather, but no less amazing in those circumstances. Who had done this, and why? Hard to imagine a more unlikely collection, unless it was the Cowdens and their actual JAL herd. Clearly there was a connection—the lead cowboy was pointing south toward Jal, once the old Cowden "Muleshoe" camp.

I drove off the highway up a dirt road through the mesquite and prickly pear to the ridgeline and out to the figures. They were enormous, cowboys on horseback maybe twenty feet tall, cows about ten feet. They were cut out of quarter-inch steel plate, mounted on heavy pipe buried in the ground. No telling how much they weighed or what it took to set them up. I was reminded of the Cadillac Ranch outside of Amarillo, an installation by Stanley Marsh, who buried twelve classic Cadillacs nose down in a pasture beside the interstate. Drivers all of a sudden pass this row of cars with fabulous tailfins, stuck like arrows into the earth. Shocking and humorous. But that's in Amarillo, a city. Jal, New Mexico, doesn't get much commuter or tourist traffic. Someone out there had a vision, someone heat-struck or inspired, the same sort of soul as the Cowden boys in 1883, deciding that this forsaken country, by God, was where they'd raise cattle. I'll have to track down the responsible party and get him help.

I headed back toward the state line to look for Monument Draw, where JAL headquarters were located in 1886. From *The Handbook of Texas* I'd learned that Old Adobe was in New Mexico some three miles west of the Texas line. Though the ranch house was gone, I knew that the Cowdens had dug wells and erected windmills; perhaps I could locate the remains of those labors. You might trash a house in the West, but never water. Where signs announced Dollarhide Road and Union Oil's West Dollarhide Field, I turned off the highway onto a sandy road, following it back past one pumpjack after another. Off to my left was the slight but noticeable draw, where I watched for trees of any sort above the mesquite flats—signs of present or past waterings, of water wells and windmills tapping into the aquifer below. Just like the Cowdens, if I found water, I might have something. When they were searching, however, there were no oil and gas wells, no roads, no clues from books. They had word of mouth, and nothing more to guide them than their experience with tough country. Still, we had a common quest.

On two or three occasions, I spotted trees and windmills and followed tire tracks through the brush to check them out. In each case, I found a windmill and watering tank, a few trees, a few cattle, but no hard evidence of habitation. Any one might have been the site of Old Adobe, or none of them. At my last stop, where I found some barbed-wire corrals and remains of a small shed, I resigned myself to uncertainty. I wouldn't find the exact location today, but I was close. This was the heart of the JAL Range, which had more to do with a vast landscape than a small adobe house. I would find the JAL everywhere and nowhere, like most open-range ranches.

Also, I saw jackrabbits running through the brush and took that as a sign. Whether back in Midland or out here, rabbits and Cowden ranches seemed synonymous. On my walk back to the car, I retrieved a tumbleweed hung in a fence—a JAL tumbleweed, I told myself. I picked up a bit of rusty barbed wire and a desiccated wand of cholla. Whether or not the rabbits and tumbleweeds, barbed wire and cactus, were actually speaking to me or I was speaking through them to myself, I felt stirrings, heard voices from the past—echoes.

So I'm back now in Midland with my tumbleweed, barbed wire, cactus, and undeveloped film, where ghosts await their appearance in good time. Jal was good to me, like it was good to the Cowdens—the day has been full of signs and wonders, which are mine to recognize. Tomorrow I hit the trail again, back east to Palo Pinto, back in time 150 years.

May 19, 2000—Today I drove from Midland to Palo Pinto, backtracking another Cowden journey. In the 1880s the "Cowden boys," sons of W. H. Cowden Sr., took their cattle herd from Palo Pinto County to Midland, more than two hundred miles across West Texas. I don't know their exact or even approximate route, though water had to be an important consideration—the only river I crossed was the Colorado. I passed Big Spring, Colorado City, and Sweetwater, their names drawn from old waterings the Cowdens may have used on the trail. Following I-20, I tried to envision the country they encountered more than a century ago, without the towns now lining the interstate. Like Coahoma, Texas, which boasts: "Home of Quail Dobbs, Rodeo Clown."

Sam suggested that mesquite has become more prevalent and widespread over time, but I'm not sure that's a fact. Mesquite is an adaptable species, like coyotes, but not new to Texas—Cabeza de Vaca observed it in the sixteenth century. Prairie fires that suppress growth are less common now, and as the country on my route gradually changed from semiarid tableland to rolling prairie, it was clear that mesquite growth changed with topography and weather—scattered low bushes near Midland evolved into thickets of trees east of Abilene. By the time I reached Strawn, in the southwestern corner of Palo Pinto County, other trees like cedars and oaks were competing with mesquite and prickly pear. Then, from a rim of rocky hills a broad valley opened up, bottomland cleared for pastures and farming— the first signs of intensive agriculture. I was no longer in West Texas.

I drove through Strawn—detouring through a cemetery on a blind, futile hunt for the grave of my great-great-grandmother—then headed for the hills to the north. The soil grew thin and rocky, cattle and goats grazing rough country covered in thickets of oak and cedar, a broken dense landscape entirely different from what I'd seen near Midland and Jal. Was this where the Cowdens spent thirty years raising families and cattle, and fighting Indians? It was hard to know what I'd eventually find out, but everything around me was charged with potential: any creek might have been their home water, any bluff the site of some battle with man or beast. Give the imagination sufficient contact with reality and the possibilities take on great vitality.

I drove into Palo Pinto, a sleepy county seat with a big sandstone courthouse surrounded by little shops: Oak Street Gallery, A Little Bit Western, Palo Pinto Cafe. I didn't stop, because I'm returning to put in a full day or two of research. I left town and crossed the Brazos River, winding through the hills, then passed through Mineral Wells, the only city of any size in the county. I headed for Fort Worth, then Dallas to stay with yet another cousin—there is no end of them! I'm now at Guy Kellogg's house in the city, which continues to rise and sprawl over the landscape, hardly evoking southwestern history, except by nostalgic effort. "Big D" residents don't quite know whether to pursue their old roots or their new growth, but for the most part, urban cowboys are all you'll find here. Neiman Marcus has supplanted western-wear stores, and Dallas Cowboys football is currently running second to the ice hockey team in the public consciousness. I'll be ready to return to actual cow country—Fort Worth, not Dallas, is the historic gateway to West Texas—and see what I can uncover about the Cowdens in Palo Pinto.

May 22, 2000—Dallas is well back from the frontier where my story lies, so today I returned to Palo Pinto. The previous year, Guy Kellogg and I spent a day there searching for anything and everything Cowden. At the courthouse, I asked to see the Marks and Brands Records, and the ladies in the county clerk's office looked at me quizzically. "What date?" one of them asked. "1850s," I said. "As far back as they go." We were sent down to the basement, where it was cool and quiet, old court records filed in tall, dimly lit stacks. Guy and I leafed through the dry pages and turned up several Cowden brands. We also located records of land sales, which I hoped would lead to the exact location where the Cowdens ran their cattle. Upstairs, the clerks made copies for us of the antique documents, written in the florid hand of the nineteenth-century clerk of court.

After leaving the courthouse, Guy and I drove to the Boyce Ditto Public Library in Mineral Wells. Boyce Ditto was undergoing renovation, so most of the genealogical records and local histories were unavailable. We found an account or two, made our copies, called it a day. My first foray to Palo Pinto had done little more than deepen the mystery of the Cowdens' time on the frontier there.

This trip was more productive. After Fort Worth, I drove to Weatherford, where Oliver Loving is buried, his body brought by wagon from New Mexico, where Indians had mortally wounded him on a trail drive. "Take me back to Texas," Loving asked his partner, Charles Goodnight. "Don't leave me in foreign soil." A biography of Goodnight by J. Evetts Haley was followed eventually by novelist Larry McMurtry's *Lonesome Dove* heroes, based on the two men. Both were early Palo Pinto cattlemen at the same time as the Cowdens, and I'm convinced that accounts brought back by their neighbors drew the Cowden family to West Texas and New Mexico. For all I know, Cowdens may have ridden the Goodnight-Loving Trail at some point.

In Weatherford, I stopped at the library, which I hoped would have historical and genealogical materials. It did, a whole room devoted to local and family history. I found tax records, military records, and, most important, a cemetery record for Palo Pinto County. I was determined to find Carrie Liddon Cowden's grave. My great-great-grandmother had twelve children, astonishing to us now. It's difficult to imagine her life of pregnancy and babies, housework and chores, all on the frontier. Undoubtedly she simply gave out. Twelve children! My great-grandfather was her last, in 1875. After her death, William Hamby Cowden, left with a full house and no helpmate, married again—to Carrie's widowed sister, Kittie. Not much different from Plains Indians who made multiple wives of sisters. I wanted to find out anything I could about the startling particulars of Carrie's life.

I found the cemetery record, in which rested "Carrie M. Cowden, wife of W. H. Cowden, b. 5-1-1832, d. 9-16-1879." Other Cowdens were listed as well, all buried in Davidson Cemetery: "The burial ground . . . has 363 graves. Oldest grave with a marker is Miles Edwards, d. 11-22-1862. Average life span was 46.2 years. Longest life span was Alice Evans Allen, 97 years. . . . Located between Thurber and Strawn. Beginning at the Mingus

Post Office, go south 2.5 miles, turn right, follow pavement 2.4 miles to Davidson Cemetery." Thurber, Strawn, Mingus, Davidson Cemetery—I had actual places to search now. To get there, I took the interstate to Gordon (population: 516), the "1996 and 1999 6-man football State Champions." Six-man football, the sport of places too small to field a full-sized team, but the only game in town for the boys and their fans. Carrie and W. H. Cowden could have fielded a team from within their own family. Two teams, if you let girls play.

The directions led me out into country carved by arroyos between low hills, dry-looking except for the mixed cedar and mesquite. Everywhere were places for wild cattle and Comanches to hide. Around a bend I came upon the cemetery, surrounded by a low red-brick wall bordered by deep-green cedars, a lovely place of rest in rough territory. At the entrance was an ornate, towered, brick-and-wrought-iron gate—I'd be entering the castle of eternity. With my cameras, I passed through to the other side and began walking the rows of weathered tombstones.

And there she was, her stone. She lay in the shadow of a spreading cedar, beside her Mary J. Bell, "dau. of W. H. & C. M. Cowden, Born 11-10-1849, Died 11-21-1868." I did the math: nineteen years old. On the other side was Nannie, "dau. of W. H. & C. M. Cowden, Born 2-25-1867, Died 11-15-1867." Nannie had lived just nine months, the only Cowden child to die as an infant. Nearby was Willie J. Jowell, "Wife of J. T. Jowell, Born 1-3-1851, Died 1-16-1880." Willie Jane was yet another Cowden daughter, yet another loss on the frontier. Still, I felt exhilaration in finding these Cowden women, together, still a family. I looked around, over the brick wall into a draw thick with mesquite. Nearby the South Fork of Palo Pinto Creek cut its way towards the North Fork; they join and flow into the Brazos and down to the Gulf of Mexico. I felt whole, transcendent, going forward by going back. Lines came to me from Whitman, who published *Leaves of Grass* about the time the Cowdens came to Palo Pinto:

It avails not, time nor place—distance avails not,
I am with you, you men and women of a generation, or ever so many
generations hence,
Just as you feel when you look on the river and sky, so I felt . . .

After my mystical moment in the country graveyard, I drove back to Palo Pinto for another session in the basement of the courthouse, looking through deeds for more signs of life. Soon I start the long drive back east, away from all this frontier adventure, but perhaps it avails not—time nor place nor distance.

Cowboys and Indians

Texas was a new country then, and certainly an aggressive country. Every bush had its thorn; every animal, reptile, or insect had its horn, tooth, or sting; every male human his revolver; and each was ready to use his weapon of defense.

Lieutenant Colonel Richard Dodge,
The Plains of the Great West and Their Inhabitants,
Being a Description of the Plains, Game, Indians, &c.
of the Great North American Desert

For anyone who loved God's own creation, it was a paradise on earth. There were springs all over the country. Streams were full of all kinds of fish, the hills and valleys were full of deer, turkey, antelope and bear and also bee trees could be found all over the country and plenty of buffalo. Everything was there to make people happy and life worth living and plenty of Indians thrown in.

W. C. Cochran, "Story of the Early Days, Indian Troubles,
and Cattle Business of Palo Pinto and Adjoining Counties"

When W. H. and Carrie Cowden arrived in Palo Pinto—probably in late spring when grass along the trail was abundant, their mule-drawn wagons packed with necessities and babies, a herd of thirty Longhorns trailing behind, pushed along by W. H. on horseback—before them lay the empty plains. They were still within reach of water and wood, on the western edge of the Cross Timbers, described in 1852 by army explorer Randolph Marcy: "It seems to have been designed as a natural barrier between civilized man and the savage, as, upon the east side, there are numerous spring-brooks, flowing over a highly prolific soil, with a superabundance of the best of timber, and an exuberant vegetation . . . while on the other side commence those barren and desolate wastes, where but few small streams greet the eye of the traveller, and these are soon swallowed up by the thirsty sands over which they flow; here but little woodland is found, except on the immediate borders of the water-courses."

Originally called Brazos de Dios—Arms of God—supposedly by grateful parties of Spanish explorers dying of thirst, from headwaters far across Texas the Brazos River now snaked its way through low, rugged limestone hills covered with cedar brakes and post-oak thickets. Villages of Wichita and other sedentary Indians lay along the river, and since they habitually set fires to flush out game, underbrush was limited; mesquite and grama and buffalo grasses flourished. Into the Brazos ran rocky feeder streams— Keechi and Rock creeks from the north, Ioni and Palo Pinto creeks from the southwest. The Cowdens headed up the Palo Pinto, where George Franklin had located a cattle range the year before. The creek gave name to the county, but the Palo Pinto, lined with "painted wood" that turned vivid colors in autumn, lay beyond most settlement. Writing a century after the Cowdens arrived, John Graves in his classic *Goodbye to a River* caught the remote nature of the area: "Nights are cool and days blue and yellow and soft of air, and in the spread abundance of even a Texas autumn the shooting and the fishing overlap and are both likely to be good. Scores of kinds of birds, huntable or pleasant to see, pause there in their migrations before the later, bitterer northers push many of them farther south. Men and women are scarce."

Scarce they were in 1855, which was the point. The two Cowden brothers had gone to Texas for apparently limitless land and found themselves crowded; now they sought out the farthest edge of the frontier. After annexation of Texas in 1846, the federal government had established a line of military posts to protect settlers from Indian raids. Fort Worth, back east eighty miles, once stood on the steadily advancing frontier, but pressure for more land led the government to move the line west more than a hundred miles. In 1851 Fort Belknap was established up the Brazos, encouraging settlement downstream. Among the daring, self-reliant pioneers were Bill and Frank Cowden, who had few neighbors in the southwestern corner of the county, where the hills gave way to rolling plains and the country was fine for cattle but little else. They settled just west of the 98th meridian, the line often used to demark the "humid," timbered East from the semiarid plains environment. Behind them the country was green; not so the mesquite and grama grasslands before them. "There were no farms there then," recalled Irbin Bell, who came as a child to southwestern Palo Pinto County, "not even a garden."

No close neighbors, no garden for Carrie to tend, but good grazing for cattle, and wild game plentiful to support a family. A stand of corn might come next year, to grind into meal at the new mill in Palo Pinto; cornbread and meat were the staples of most frontier families. Some brought hogs or milk cows. Bill Cowden went to work settling in, the family probably living in a tent while he built a "dog run" cabin—two small, separate rooms under one roof with an open breezeway, or dog run, in between. Of cedar logs chinked with clay, with a front porch across the length of the structure and a rock chimney at each end, the Cowden home was sufficient to the family's needs, little more. A clapboard roof kept the weather out, off simple furniture, off pallets laid down on hard-packed dirt floors. Perhaps there was glass in the windows, more likely just shutters that closed from inside, for protection. Traveling through Texas at the time, Frederick Law Olmstead stayed in similar cabins, comfortable enough, though he could "look out, as usual, at the stars between the logs."

The Cowdens had it better than most, however; they had family nearby to help with cattle herding and other labor. As usual, the brothers worked in tandem, taking care of stock and family. Frank's family had probably remained back in Shelby County with W. H. while Frank "broke trail" out to the frontier, but now the Cowdens were all together, which helped relieve the sense of profound isolation frontier folk could suffer. "The enormous strength of this breed lay in their complete rejection of the organism of human society," proposed Fehrenbach in *Lone Star*. "The Texan, like all Westerners, was not antisocial, but asocial. He congregated or cooperated only for education, or defense, and then with some reluctance. No other breed, probably, could have lived contently for years on a far-flung frontier, where the distances between houses or farms was measured in miles." Under such conditions, "society" usually meant one's family—cohesive by necessity, a compact version of the Scotch-Irish clans back in the Appalachian Mountains or Indian bands roaming over the plains. Though small, the Cowden clan was doubly strong.

Like them, others moved into the country looking for grazing land. In Keechi Valley north of the Brazos, Oliver Loving and young Charles Goodnight ran herds; the Slaughters bought land along the river. In *C. C. Slaughter,* David Murrah wrote: "The Slaughters' choice of a new range was not difficult. . . . [T]he Brazos valley provided a broken, well-watered terrain; tough mountain cedar, Spanish oak, cottonwood, ash, and pecan trees offered ample material for housing and fencing; nearby Fort Belknap promised military security. In addition, two nearby Indian reservations, the Brazos Reservation for the Wichita and other East Texas sedentary tribes and the Comanche Reservation on the Clear Fork of the Brazos, offered a ready market for cattle. Also, other settlers along the Brazos—the Cowdens, Daltons, Goodnights, and Lovings—shared the Slaughters' enthusiasm for cattle raising." These cattlemen helped establish Palo Pinto as a major wellspring of the southwestern cattle kingdom to come, as Joseph Carroll McConnell asserted in his modestly titled *The West Texas Frontier, or A Descriptive History of Early Times in Western Texas Containing an Accurate Account of Much Hitherto Unpublished History, Presenting for the First Time in Historic Form a Detailed Description of Old Forts, Indian Fights and Depredations, Indian Reservations, French and Spanish Activities, and Many Other Interesting Things:* "Unquestionably, Palo Pinto County has produced more noted cowmen than any of the frontier counties." Some called it the "cradle of cowmen."

At the time, there was no sense of collective destiny, just constant hard work. In addition to tending his own small herd, W. H. helped neighbors with their cattle, keeping them from drifting off into the rough hills or endless prairie and joining "cow hunts" to bring them back. He was horseback much of every day, all week, year-round. Carrie tended the house, the five children—John Motherwell was born in 1858—and countless backbreaking chores. The Cowdens occasionally traveled fifteen miles to the newly organized county seat. First called Golcando, "Palo Pinto Town" had a new courthouse, Dick Jowell's blacksmith shop, and Captain Dillahunty's store with news from near and far, a barrel of whiskey for paying customers, and goods for the ranches scattered throughout the area. "The men looked to him [Dillahunty] to keep the company supplied with powder and lead," wrote Mary Whatley Clarke in *The Palo Pinto Story*. "He carried shoes, hats, gloves, domestic, calico, sugar, salt, molasses, soda, coffee and whiskey, and many other things." One rancher's bill included "saddle, $21.00; four pounds grass ropes, $1.37; water bucket 60c, pair lady's cotton hose 35c, set knives and spoons $2.50, set cups and saucers 50c, pair cotton cards 50c, one lead pencil 10c, hatchet $1.00, and 13 fish hooks 10c." A quart of whiskey cost 50 cents; four plugs of tobacco ran $1; a churn brought $2.50, an iron cookstove a whopping $43. Most goods were freighted by ox-drawn wagons from Galveston down on the Gulf—Charles Goodnight was a teamster from 1857 to 1860—and most settlers ran up bills for months. Paydays and currency were both scarce. "When a customer died, owing a bill," Clarke noted, "Captain Dillahunty often marked it off, and wrote 'Paid by God.' He did not hold the widow responsible for the debt."

Customers died from natural causes, although they were generally a healthy lot, isolated from diseases that afflicted cities along the Texas coast. More often the cause of death was neither natural nor accidental, since Palo Pinto in its early days was not only the frontier cradle of cowmen—it was also ground zero in the Texas Indian wars.

Summers when we traveled to Texas and New Mexico, I crossed into the world of cowboys and Indians—a real world inhabited by imaginary people. Indivisible parts of one thrilling whole, cowboys and Indians rode around in my head, shooting rifles or bows and arrows, romantic figures shaped by *How the West Was Won* and *The Roy Rogers Show* and *Broken Arrow*. But cowboys and Indians were also shaped by direct contact with the actual landscape of ranchers and Comanches and Apaches. Mesas from which warriors sent ominous smoke signals in movies appeared before my very eyes as we drove Route 66 from Amarillo to Tucumcari to Santa Rosa. In the backseat of our Ford station wagon I felt in peril, blood quickening as I scouted rimrocks for flashes of sunlight off rifle barrels. The failure of Indians to appear seemed only to confirm that they were there.

On arriving in Santa Rosa we went first to the Club Cafe, plugged on billboard after billboard across West Texas with a happy, plump man gobbling a huge steak. There we'd find my grandparents waiting—Big Guy in his white Stetson, little Annie Mae beside him—to have iced tea and a meal. Much as I loved seeing them, I could hardly wait to scramble from the dining room and go wander the aisles of the curio shop attached to the cafe. I'd check out the cheap trinkets—gaudy war bonnets, little bows and arrows, tomahawks of plastic and bamboo—that proved we had crossed into Indian territory. Each year I grew into a new pair of cowboy boots and a new pair of moccasins.

Once we left Santa Rosa, off on the dirt ranch road, my transformation from car passenger to pioneer was complete. I could see no signs of civilization other than an occasional barbed-wire fence disappearing across the open plains. We were alone out there . . . or not. Over each rise an ambush might await us! With lances and shields, Comanches on painted war ponies lined up to ride us down without mercy. It was cactus and rattlesnakes and Indians the whole way to the ranch. No wonder we were always so excited to spy the ranch tucked back against the mesa—we had survived!

Within the safety of its walls were yet more reminders of transport into another world, a world of confrontation with fierce natural forces—a rug made from the hide of a mountain lion, snarling head still attached; rifles stacked on a rack above the fireplace; a collection of diamondback rattles on a mantle. And what were Indians but another force of nature? At night when coyotes howled—or were they coyotes?—I spent hours studying a book of Indian sign language, photographs of Chief Iron Eyes Cody demonstrating signs for antelope, buffalo, friend, horse, moon, rain, tepee, war. I learned how to say *To-travel-through-an-unknown-country* and *I-knocked-him-over-with-a-single-shot* and *He-reached-here-yesterday-came-into-my-lodge-we-talked*. My dreams at night, when I finally fell asleep, carried me even further into primitive, dangerous, stimulating country. In that I was probably little different from a child on the actual frontier.

But of course I awakened. And eventually came to distinguish fantasy from fact, though both remain real. I've taught Native American studies and now know how distorted and romanticized were the Hollywood cowboys and Indians of my childhood—life wasn't as simple as red and white. I've learned that the Indian wars were effectively over before 1883, when the Cowdens moved to New Mexico, although Geronimo did not finally surrender until 1886. On the other hand, I've also learned that the original Cowdens back in Palo Pinto County faced terrors and dangers equal to anything I imagined as a child. The challenge is to imagine their extraordinary reality rather than some romantic notion—the truth is more than enough.

Walter C. Cochran remembers:

> The early settlers on the Texas frontier were a happy and contented set of people until the Indians went on the war path in the fall of 1859. Jack, Young, Palo Pinto, and Parker were the worst Indian counties in Texas. There have been more men killed and women and children captured and taken off prisoners in Jack County than any other county in Texas. It was close to the line of Red River, and they had no trouble in getting away with their prisoners and horses.
>
> The first raid the Indians made through Palo Pinto, Erath, and Hood counties after they went on the war path was in the fall of 1859. There were not many horses in the country at that time. My father had three horses; they stole two of them; they got four from Frank and W. H. Cowden and all of Dick Lloyd's, and part of what Peter Davidson and Will Allen had. They got a few horses from all the other ranches, then went to Erath and Hood counties, stole forty or fifty head, and then went back to the reservation at Fort Belknap.

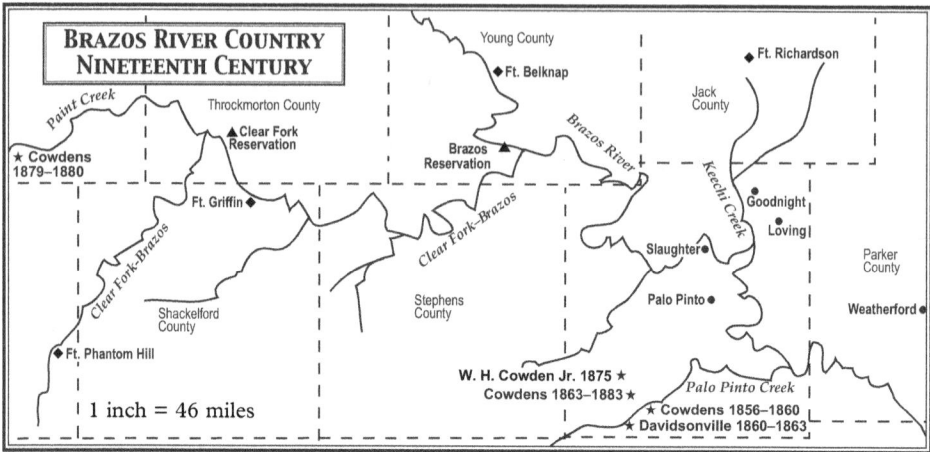

BRAZOS RIVER COUNTRY NINETEENTH CENTURY

Paint Creek

★ Cowdens 1879–1880

Throckmorton County

▲ Clear Fork Reservation

Ft. Griffin ◆

Clear Fork-Brazos

Shackelford County

◆ Ft. Phantom Hill

1 inch = 46 miles

Young County

◆ Ft. Belknap

Brazos Reservation ▲

Clear Fork-Brazos

Stephens County

Brazos River

W. H. Cowden Jr. 1875 ★
Cowdens 1863–1883 ★

Jack County

◆ Ft. Richardson

Keechi Creek

Goodnight ●
Loving ●

Slaughter ●

Palo Pinto ●

Parker County

Weatherford ●

Palo Pinto Creek

★ Cowdens 1856–1860
★ Davidsonville 1860–1863

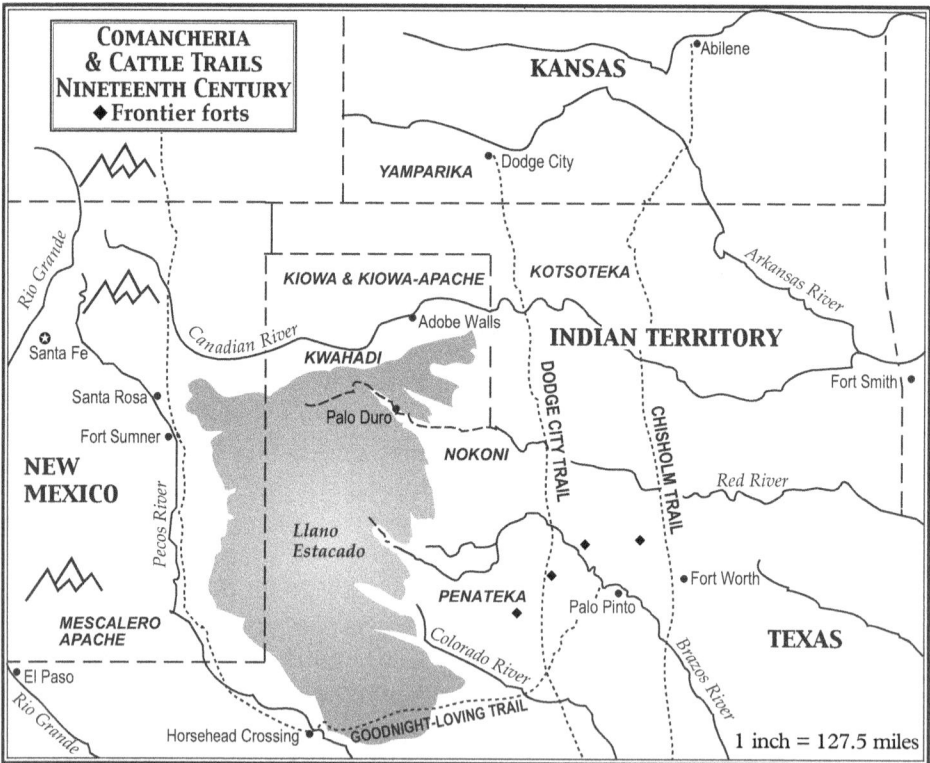

COMANCHERIA & CATTLE TRAILS NINETEENTH CENTURY
◆ Frontier forts

KANSAS

● Abilene

YAMPARIKA

● Dodge City

Rio Grande

KIOWA & KIOWA-APACHE

KOTSOTEKA

Arkansas River

Canadian River

Adobe Walls ●

INDIAN TERRITORY

✿ Santa Fe

KWAHADI

Santa Rosa ●

Palo Duro ●

Fort Smith ●

Fort Sumner ●

NOKONI

DODGE CITY TRAIL

CHISHOLM TRAIL

Red River

NEW MEXICO

Pecos River

Llano Estacado

◆

◆

● Fort Worth

PENATEKA

◆

Palo Pinto ●

MESCALERO APACHE

Colorado River

Brazos River

TEXAS

● El Paso

Rio Grande

Horsehead Crossing ●

GOODNIGHT-LOVING TRAIL

1 inch = 127.5 miles

Cochran's account is one of hundreds that describe conflicts between Texans and Indians, from the Texans' point of view, which most often ignored Indian perspectives, like that offered by Comanche chief Ten Bears in Dee Brown's *Bury My Heart at Wounded Knee:* "My people have never first drawn a bow or fired a gun against the whites. . . . So it was in Texas. They made sorrow come in our camps, and we went out like buffalo bulls when their cows are attacked. When we found them we killed them, and their scalps hang in our lodges. The Comanches are not weak and blind, like the pups of a dog when seven sleeps old. They are strong and farsighted, like grown horses. . . . If the Texans had kept out of my country, there might have been peace. But that which you now say we must live on is too small. The Texans have taken away the places where the grass grew the thickest and the timber was the best. . . . The white man has the country which we loved." More typical of Anglo perspective was J. W. Wilbarger's 1889 *Indian Depredations in Texas: Reliable Accounts of Battles, Wars, Adventures, Forays, Murders, Massacres, etc., etc., Together with Biographical Sketches of Many of the Most Noted Indian Fighters and Frontiersmen of Texas,* which runs 691 pages of gory horrors—isolated events hard for participants to see as rooted in culture and history. The steady, bloody expansion of European, then American, territory westward from the Atlantic coast had the effect over three centuries of creating a series of "last stands" by native tribes. Custer's stand on the Little Bighorn in 1876 was actually the last stand of the Sioux and other northern Plains Indians, crowded into smaller and smaller territory, like cattle rounded up by cowboys. Tribes first fought one another—Dakota against Crow, Comanche against Apache—for hunting grounds and horses, then fought whites overrunning their land. "As in the case of most American Indians in historic times," Wallace and Hoebel observed in *The Comanches,* "war was the principal occupation of the Comanches," who battled competing tribes of Utes, Apaches, and Pawnees, then the Spanish, Mexicans, Texans, and Americans, as each in turn intruded upon their territory. "They made life unsafe for their enemies who dared to live near the periphery of the Comanchería. Throughout the Southwest their name has become a synonym for wildness, fierceness, and savagery."

To daring settlers like the Cowdens and others in northwest Texas, the Comanchería—horse warriors inhabiting a hostile environment— was for years an elemental barrier to any advance. Unlike tribes in eastern Texas—Caddos, Wichitas, and others crushed by troops of the Texas republic—the Comanches could be pacified by neither force nor concili- ation, in part because the tribe was never settled and never unified. The scattered Comanche bands included the Penateka (Honey Eaters), Nokoni (Wanderers), Kwahadi (Antelopes), Kotsoteka (Buffalo Eaters), and Yamparika (Root Eaters); Kiowa and Kiowa Apache allies at times joined one or another band for raids. Treaty agreements with one band were unknown or unrecognized by another, and the various bands roamed such vast and inaccessible terrain that conventional military maneuvers had few points of attack. Texans living on the frontier had a fluid, ghostly foe; it's not surprising that most white settlers came to treat all Indians with suspicion, hostility, and outright aggression.

Comanche culture, like that of acquisitive and aggressive Anglos, also helped to ensure conflict. The nomadic life of the Comanches laid claim to widespread and inexact territory, within which individuals were free to pursue their own ambitions, and the foundations for social recognition were war and plunder. Colonel Richard Dodge wrote in *The Plains of the Great West*, "The grandest of exploits and the noblest of virtues to the Indian are comprehended in the English words—theft, pillage, rapine, and murder. He can expect no honour from man, or love from woman, until he has taken a scalp, or at least stolen a horse." Comanches had been raiding Mexican settlements for decades, stealing horses, capturing women and children, following the dictates of their warrior culture, although none of this was well understood by new Texans. The results were inevitably disastrous, as Wallace and Hoebel noted: "Gradually the Anglo-American settlers in Texas scattered fearlessly over the face of the country; each one built his cabin and plowed his field wherever his fancy dictated, without much regard to means of defense beyond his individual arms and without apparent desire for near neighbors. They brought with them a superior breed of horses, much larger, stronger, and more enduring than the neglected breed which had grown from Spanish stock. Such a situation made it easy, attractive, and relatively safe for the Indians to take horses and scalps for trophies. It invited raids."

Comanche and Kiowa raids stopped the headlong advance of the frontier in Texas for decades and even briefly forced its retreat. However fearless or foolish initially, whites in Palo Pinto County in 1860 came to appreciate the danger of their position on the threshold of the Comanchería. Every moonlit night seemed to bring another round of raids upon the isolated cabins; parents put their children to bed, fearing for their lives. Even though "a great many of the first settlers of the Texas frontier were Mexican War veterans and they liked excitement and trouble," W. C. Cochran remembered incidents that horrified the whole frontier:

The women folk were left alone at the houses while the men went for wood, supplies, etc. It was one of these occasions . . . Indians charged one unprotected home and carried four helpless women off. The home raided was that of Mrs. Wood, a pretty woman who had the three Lemley girls who lived near by for company while her husband was away. The youngest Lemley girl was fourteen and the oldest eighteen. The girls, never dreaming of danger, were having a good time together when their blood seemed to freeze in their bodies as the war whoops were heard about the small log cabin. . . .

"It was no time until the Indians had rushed into the cabin, grabbed each terrified girl, and swinging them to their horses were away as fast as they could run. About three miles from the cabin, the Indians became disgusted with the youngest Lemley girl for weeping; so they stopped their horses and pulled the girl to the ground. Here they found pleasure in torturing the crying girl with a large knife, waving it before her face and threatening to cut her throat. They actually pressed the blade so hard against her tender flesh that the skin was cut and badly bruised. The heavy dark hair of Mrs. Wood held their fascination just then more than throat cutting, and the young girl was left alone while Mrs. Woods' lovely hair was cut and jerked off while the poor woman was still alive. After the savages tortured Mrs. Wood and the middle Lemley girl, both were killed and their dead bodies brutally assaulted. The other two girls expected the same fate as they were carried on with the Indians the rest of the day and the following night, but the next morning after stripping off all their clothes except petticoats, shoes, and stockings, they were turned loose.

Parties of Nokoni and Penateka Comanches—"wild" Indians joined by others from the reservation on the Clear Fork, northwest of Palo Pinto—were responsible, or at least blamed, for many raids. Facing angry opposition by frontier counties, Texas reservations were closed by 1859, though removal of Comanches and other tribes to Indian Territory did not stop depredations. Cycles of Indian raids and retaliatory campaigns by Texas Rangers continued with little relief for frontier counties. Word came to Dillahunty's store in Palo Pinto of pillage, rape, and murder: "The agonizing screams of the victim seemed to delight the heartless monsters, and it was not until they had inflicted upon this poor woman every character of punishment which their devilish minds could invent, that they could make up their minds to leave. . . . They deliberately stripped her of all her clothing, shot several arrows into her body, and when ready to leave, two Indians on horseback rode up on either side and each taking hold of her, dashed off, while a third Indian followed behind and beat her in the back with a heavy stick. Finally, she fell almost lifeless upon the ground, when an Indian warrior dismounted, passed his knife around her head, and tore off her scalp." The Mason and Cameron families, the Hale and Catbey boys, John Brown, Joseph and Frank Browning—the roll call of murdered Palo Pinto settlers grew. At the same time, innocent Indians died at the hands of whites; in 1858 a party of peaceful Caddos in Palo Pinto County was attacked while sleeping, with seven killed. The killers were from adjacent Erath County, but Palo Pinto residents had to brace for a reprisal. Elements on both sides were striking at any convenient target.

Comanche activity intensified after the Civil War began in April 1861. "The people on the Texas frontier woke up and found themselves with two wars to fight," remembered Cochran. Federal military posts, including Fort Belknap northwest of Palo Pinto, were abandoned by Union forces at the same time many Texas men were called away to serve the Confederacy, leaving the frontier but lightly protected. A dark time lay ahead.

The two Cowden brothers had served as volunteers in the Mexican War and perhaps wanted no part of another conflict far from home, especially with large families to protect. When citizens gathered at the Palo Pinto courthouse to draw lots for either the Confederate or Frontier regiment, W. H. and G. F. Cowden drew duty in Texas, serving in Company A of the First Frontier District. "All the men on the frontier of Texas that did not go to the war joined the rangers for the purpose of guarding the Texas frontier. They were required to range ten days out of each month and more if called upon to serve," recalled Cochran. "The balance of their time they had their hands full taking care of their families, trying to keep the Indians from murdering and carrying them off as prisoners for the ransom they might get for them." Many settlers retreated east, to towns back from the frontier. Cochran described how remaining settlers survived: "The Indians kept getting worse, and people began forting up all over the country. Four to a dozen families would fort up in neighborhoods. The Allens, the Stuarts, the Strawns, the Cowdens, and several others forted up at Peter Davidson's ranch about four miles southeast of where Strawn now stands." The neighbors built picket houses of upright logs and covered them with dirt, and men not on ranger duty ventured out by day to do what work they could. Irbin Bell also remembered life at the frontier forts: "They would build a schoolhouse and would have a guard to go with the children to school, a well-armed one. Two men guarded the children to school one year. We were with the group on Palo Pinto Creek at Peter Davidson's ranch when they forted up in 1860. There were about a dozen families there, and the place was called the Davidson School House. . . . Frank Cowden, Bill Cowden Sr., Peter Davidson, Bethel Strawn, Jack Stuart, Rube Clayton Sr., and a man named Martin whose youngest son was killed by the Indians all built in there together. Theodore Wright was the teacher." Like teachers a century later, drilling students on what to do if atomic bombs began to fall, Theodore Wright surely had instructions for his frontier children in case of a Comanche attack: duck and cover.

Life went on despite Civil War and Indian raids. "Tough times don't last," goes the cowboy wisdom, "but tough people do." Times and people both were tough during the war, as Union blockades closed the Mississippi River and the Gulf of Mexico by 1863. Goods could get neither in nor out; tools, ammunition, clothing, sugar, coffee, medicine—these and other items were hard to come by. Writing to state officials, Palo Pinto judge W. L. Lasater pleaded for scarce but critical supplies: "There are at least 160 families in this county that should have at least one pound of powder and lead and caps in proportion to each family, not knowing at what hour they may be attacked by the merciless savages."

"In 1863," wrote Cochran, "Frank and W. H. Cowden left Davidsonville and settled on Rock Creek about eight miles northeast of Strawn. Henry Bradford and Green Davis built some picket houses and lived with the Cowdens till after the war closed." While forted up on Rock Creek, the Cowdens also endured drought conditions in 1863, when even the Brazos quit running. And that winter a series of nasty northers swept down and struck the country, prompting cattle from all over north Texas to drift south before the storms, sometimes to the Gulf of Mexico, well beyond the reach of owners busy fighting Indians or Yankees.

Tough times, yet men nor women seemed deterred. Carrie did her part, delivering three more children—Charles Webster in 1860, Cynthia Katherine in 1862, Annie Lee in 1864—bringing the number to eight, the eldest only fifteen. As the family grew, so did their cattle herd, though it did little good at the time. There were limited markets for beef before the Civil War and almost none after Union blockades were in place; New Orleans, once the nearest major market for Texas cattle, was occupied by Union forces. There were a few other prospects. Oliver Loving had trailed cattle from Palo Pinto to both Illinois and Colorado before the war; W. C. McAdams drove a small herd to Mexico, trading them for sugar and other supplies; Shreveport offered a market. But buyers were scarce, in part because money itself was scarce. Confederate notes were worthless, used as drawing paper by children.

Cattle rarely sold, and the shortage of manpower meant that few bulls were castrated, so even more calves arrived each year. Throughout Texas, cattle numbers rose dramatically, as in previous decades. Census estimates in 1850 were 330,000 head; by 1860 the estimate was over 3,500,000. "Each year, up to the time of the Civil War," wrote Webb in *The Great Plains*, "cattle were becoming more numerous and less valuable." The trend continued, and at the close of the war, the price per head was only two to four dollars, if a buyer could be found. In a war-torn economy, the Cowdens and other Palo Pinto cowmen needed to find new markets for their cattle, but those markets were a long way off, and their cattle were scattered everywhere.

Samuel Maverick, a lawyer in southern Texas before the Mexican War, in payment of a debt took four hundred head of cattle running free on Matagorda Island. Maverick's caretaker "neglected to brand the calves," according to *The Handbook of Texas*. "Neighbors referred to wandering unbranded calves as 'one of Maverick's'; the phrase was shortened to Maverick and gradually came to mean any unbranded stock. In the period immediately after the Civil War when cattle herds were greatly increased, the word passed from local into general use. As long as wild Longhorns roamed the range, the appropriation of unbranded cattle was a legitimate occupation and accounted for the beginning and increasing of many herds."

When the Civil War ended in 1865, the Cowdens and veterans returning to Palo Pinto attempted to restore their herds to some order. For four years calves had gone unbranded, and herds had multiplied and mixed as cattle scattered all over the country. Like the men who tended them, Texas Longhorns were by nature a hearty, adaptable, self-reliant breed, with high fertility and birth rates; after their arrival with Spanish conquistadors, they thrived on the open ranges of Mexico and Texas. The same qualities that helped them survive made them difficult to handle and often led them to go wild. "The domestic cattle of Texas, miscalled tame, are fifty times more dangerous to footmen than the fiercest buffalo," wrote Colonel Richard Dodge in *The Plains of the Great West*. He observed the "ferocious disposition" of Longhorns, especially "the domestic animal run wild, changed in some of his habits and characteristics by many generations of freedom and self-care." Historian J. Frank Dobie described the breed in hyperbolic, "Texas" terms: "With their steel hoofs, their long legs, their staglike muscles, their thick skins, their powerful horns, they could walk the roughest ground, cross the widest deserts, climb the highest mountains, swim the widest rivers, fight off the fiercest bands of wolves . . . rustle in drouth or snow, smell out pasturage leagues away, live—without talking about the matter—like true captains of their own souls and bodies." Such were Longhorn cattle scattered over the rough terrain of Palo Pinto county, lurking in cedar thickets, alert as deer, "captains of their own souls and bodies," owned by no man—prompting the mavericking days to follow.

"It was not considered stealing at that time," W. C. Cochran recalled, "so every man that could get together a little bunch of horses—one horse to the man was enough in those days—hired a few men and went to work rounding up and branding everything they could get and kept all they got. I have known many a man that made his start in the cow business and laid the foundation for his afterwards big fortune from this mavericking." The Cowdens were among those hunting unbranded cattle, although Indian raids left them short of horses and they "commenced their mavericking days barefooted and afoot," wrote Cochran. "They would round up a bunch of old poor cows and calves, run them down and what they could not pen they would catch with their dogs, old Buck and Tige. They always put Jeff [Cowden] in the lead as he could outrun any of the rest of them. They would round up, catch, and brand ten to fifteen mavericks a day." So many men in Palo Pinto County were mavericking— sometimes on one another's ranges at the same time—that even calves sucking branded cows were not secure. Palo Pinto cowman Charles Goodnight complained about the breakdown in the frontier code: "There had been no stealing upon the northwest frontier so far as was known, and I think it was the general custom—an unwritten law—to mark and brand every calf in your range to its owner, if you knew him. If the mother cows were strays, or unknown, you branded the calves in the same brand that the cows wore . . . but certain scattered men over the country could not withstand the temptation and went to branding these cattle for them- selves." Cowmen in Palo Pinto eventually tried to impose some restraint, Cochran recalled: "After the work was over in the fall, men would meet and set a time to commence mavericking in the spring. There were always a few men that never quit. They would ride the country with their dogs, catch and brand all the winter." In addition to "sooners" who mavericked early, cowboys hired to gather and brand were sometimes paid by the head or given a cut of the total take, practices that undermined the rights of legitimate owners: cattle were branded and hustled away; known strays were branded; calves were branded no matter what marks their mothers carried. And Indian raids on stock continued; estimates of cattle stolen by Kwahadi and other raiders, for sale to comanchero traders in New Mexico, ranged up to 300,000 head. The whole frontier cattle business was up for grabs.

Between the two families and neighbors at their fort, the Cowdens had hands enough to gather their strays—W. H. had three sons old enough to ride—but horses were hard to come by and hard to keep from Comanche raiders, who were sometimes out for blood but most often wanted horses. Rupert Richardson noted in *The Comanche Barrier to South Plains Settlement:* "The horse was both a form of capital and a medium of exchange for the Comanche. In fact, during the nineteenth century, the horse, next to the buffalo, was the principal food for some of the bands. Horse herds could easily be transported, the Indians could find a market for horses of good quality, and they were always eager to add to their herds. Good horses were most easily obtained in the settlements of the frontier white people, and most Comanche raids were primarily for the purpose of securing these animals. If, on these forays, the braves were able to take a few scalps, the expedition was regarded as all the more successful; but the taking of human life was ordinarily a secondary matter."

Not to the Cowdens and their neighbors. But being afoot on the frontier might itself mean death, and it certainly threatened their livelihood. As for their Comanche adversaries, horses were critical. "My father loaned Uncle Billie Cowden two horses, one little gray horse, the other a sorrel mare, to maverick on. Uncle Billie did not get to use them only about a month when the Indians stole them," reported Cochran. To avoid theft, "everybody got up their horses before night, and after dark they would take them off a mile or two from the ranch and hide them out in a thicket or canyon," Cochran wrote. "I have walked hundreds of miles coming back from hiding horses and going after them in the morning. . . . I have known very few cases where the Indians would follow and get the horses after they were hobbled out. The Indians would signal one another by howling like a coyote or hooting like an owl. I never hear a hoot owl to this day but what it scares me." Irbin Bell remembered the same harrowing time and routine: "Every moonshiney night we would keep our horses hitched near the house until ten or twelve o'clock. Then the boys would take them out and hobble them. Before daylight next morning we would have to go and get them and bring them back. Sometimes the boys would take their blankets and sleep near them, bringing them back by daylight. And that was the way we lived."

Settlers were wise to guard their horses, since Comanches were renowned among all Indians, as Dodge noted: "Where all are such magnificent thieves, it is difficult to decide which of the plains tribes deserves the palm for stealing. The Indians themselves give it to the Comanches, whose designation in the sign language of the plains is a forward, wriggling motion of the forefinger, signifying a snake, and indicating the silent stealth of that tribe. This is true of the Comanches, who for crawling into a camp, cutting hobbles and lariat ropes, and getting off with animals undiscovered, are unsurpassed and unsurpassable." Cochran also testified to that skill: "My father and I went to spend the night with W. H. Cowden. The Indians were awfully bad. My father had two of the biggest bulldogs that I ever saw, old Watch and Joe. There was not a white man living that could ride up to our house and get off his horse unless someone was there to keep the dogs off of him. That night when we went to bed, we tied our horses at the door of the Cowdens' house and made a pallet for the dogs right in the door. The Indians slipped up and cut the ropes of both horses and got away with both of them and those two dogs never made a move or sound. The Cowdens had their horses hid out when we got there that night, so the Indians never got any of them. My father and I borrowed two horses and rode home. When we got there, we found that the Indians had stolen part of our horses that we had left at home the evening before."

Despite the return of Texas veterans, raids continued to plague frontier settlers, said Richardson: "After the collapse of the Confederacy, the state organization for frontier defense gradually ceased operation, and for several months thereafter the western settlers were left without protection. The Indians, soon aware of the situation, scourged the frontier as never before in history, driving the line of settlements eastward in some places by a hundred miles. . . . The worst raids were made on moonlit nights, and the soft summer moon became a harbinger of death. Charred rock chimneys stood guard like weird sentries, symbolizing the blasted hopes of pioneers and often marking their nearby graves."

Every frontier family had brushes with death or, equally painful, losses of children. Irb Bell remembered the ever-present threat to families: "I went with two six-shooters on all the time. I put them on of a morning when I got up, same as I did my shoes. When I went alone, I carried a rifle too. . . . I have tied my boy, Lee, to a bedpost in the early days to keep him from running out and playing, as I was afraid the Indians would pick him up." Bill and Carrie Cowden had eight children to protect, and in 1867 the danger struck close to home, as Cochran related: "Captain Frank Cowden and a man by the name of Winfield Scott had gone to Weatherford to mill, and on their way back home they camped near where Santo now stands. They had Fred and Jeff Cowden with them—they were mere boys. Scott had gone after the oxen and Captain Cowden had left Fred and Jeff at the wagon and went to hunt his horse when he met seven Indians coming toward the wagon with his horse. He then ran and beat the Indians to his wagon—the Indians shot at him four times before he got to the wagon. There were several shots fired on both sides before the Indians left. No one was hurt in this fight." Cochran recalled another half-real, half-imagined scare that same year: "My father and I took the horses that we had borrowed from W. H. Cowden back to them; we stayed there all night and next morning the boys went to the cow lot to milk. John M. was a big old boy then and was standing up against the fence about half asleep when one of the calves in the lot got hold of his old ducking shirt and was chewing it through the fence. Just then the milk-pen calves outside came running up to the lot with three or four arrows sticking in them. John M. started to run to the house and when the calf on the inside chewing on his shirt commenced pulling back, he started yelling that the Indians had him."

Such acute terror and chronic tension seems insupportable to modern sensibilities, but Fehrenbach pointed out: "The frontier family was functional, and maturity came earlier than it would probably ever arrive again. Work, hunger, danger, and terror could not be kept or disguised from young people. It was impossible for a boy or girl to create a false, or romantic, vision of the world." Given the harsh realities of life, it's a wonder that the Cowdens kept persevering. In February 1867, Carrie gave birth to her ninth child, Nannie Liddon, who lived just nine months—the first and only infant death in the family, indicating a remarkable rate of survival given the conditions. The next year their eldest child, Josephine, not long married, died at nineteen. A rocky run of luck had struck the family, but as cowboys say, "When you get to the end of your rope, tie a knot and hang on."

If the end of Civil War did not stop hostilities on the Texas frontier, it did open possibilities for cattlemen previously unable to market their stock. The glut of Longhorns in Texas corresponded to beef-hungry markets in the North, where cattle might bring ten times their price in Texas. Herds began moving up from southern Texas, creating trails to rail lines in Missouri and Kansas. The legendary era of trail drives—cowboys herding half-wild Longhorns and battling Indians, stampedes, swollen rivers, and hostile farmers to get their cattle to market—had begun. It would last just twenty years.

Because of its rough terrain, Palo Pinto was not directly on the Chisholm Trail, which passed to the east, nor the Dodge City Trail, which later passed to the west. Area ranchers nevertheless sent many cattle north to Kansas; the Reverend G. W. Slaughter and his son were notable drovers, taking almost thirteen thousand head up the Chisholm Trail from 1868 to 1875.

Over that period the Slaughters brought home $464,000, the start of an eventual cattle empire. J. H. Baker kept a diary of one trail drive from the Slaughter ranch in Palo Pinto to Abilene, Kansas, in 1869, a trip that lasted over two months, with its share of typical hardships and dangers. Among his daily entries are the following:

Sept. Thurs. 2; The herd of about 1230 head, (430 mine) left the Slaughter ranch for Kansas, starting in the direction of flat top mountain. J. R. Jowell, P. E. Slaughter and myself took the tally, came to Jacksboro and had it recorded. We then came out west of town a few miles and camped.

Sept. Tues. 7; It is clear this morning. Cattle very restless and troublesome during the night. Kept all hands busy. It was cold. They were very hard to manage this morning until we came to a stream where they got water, in the vicinity of Victoria Peak where we made Camp No. 6. 10 miles.

Sept. Thurs. 9; Drove to within 2 miles of Red River Station and struck camp 8. I have a sore eye.

Sept. Fri. 10; About 10 A.M. we reached the river and swam our herd across. In crossing one of our hands, Wm. Cowden came very near drowning, and it was with difficulty he was rescued. We crossed the wagons on a ferry. I crossed with them. We made the crossing without losing any of the herd and made camp 9 about 3 miles from the river. My eye has been very painful for 2 days. Ate nothing today but a small piece of bread, an onion and 4 pills for supper. Day pleasant.

Sept. Mon 13; It rained a tremendous rain last night and was so dark we could not see the cattle, so were compeled to turn them loose. Set out early this morning to find and gather them up. Spent the whole day at this job. We think we have most of them, but some are gone. Some were picked up 10 miles from camp. The ground is very wet and boggy. Had no time to move camp today. Weather pleasant after the early morning.

Oct. Mon. 5; Started early and traveled 6 miles in a N. Westerly direction, . . . having a creek on our right with some timber along its banks. We have crossed several muddy tributaries to the creek. Lands generally poor. Are grazing our horses on a patch of mesquite [grama] grass while waiting for dinner. Wm. Cowden is chilling. Traveled about 10 miles this P.M. Still up grade and N.W. Made camp 32 near the head of Clear Creek.

Oct. Sun. 10; Grazed our cattle until camp 38 was found and brought the herd to it, a distance of 3 miles, where we spent the day letting the cattle graze. The Boss went back to town. One of the hands bought a span of mules, paying for them with cows and 2 year olds to the amount of $300.00. Day hot. Mr. Cowden who has been chilling is about well.

Oct. Mon. 11; We are grazing for the cattle to be cut out in payment for the mules. By the time we were ready to start some buyers came along and bought the rest of our 2 year olds, 85 head at $12.00 each, and 40 three year olds at $17.00 each. It took us all day to cut them out, and it was after dark before we got our dinners. A cool norther tonight. Drove about 3 miles and made camp 39 on the creek about the place where we got dinner yesterday.

Oct. Tues. 19; The wind continued to blow and the cold to increase until nearly daybreak. The sleet then commenced to fall and continued for several hours. Cold and gloomy. Started early and after going a few miles met C. C. Slaughter. He is just back from Abilene, Kansas and reports favorably on the markets there. Traveled 7 miles up the creek and struck camp 46 in fine mesquite grass. The wind has ceased to blow and the weather has moderated. Are on a high prairie and have to haul our wood.

Oct. Sat. 23; The wind subsided and there is a tremendous front this morning. Very cold. The sun is beginning to melt the snow and we hope for better weather. We find many herds here waiting for buyers, many of them in poor condition, and quite shabby. Traveled 15 miles and arrived at Smoky River about sunset, 2 1/2 miles from Abilene. It has been a very cold day. We learn that the cold is bringing a rush of cattle on the market and is bringing down the price. Camp 50.

Oct. Wed. 27; Two of us held the herd last night and I am holding it today. Cold west wind. Drove the 304 to town and shipped them. Holding the balance tonight. The prairie is on fire and burning all round us, and considerable damage is being done in this vicinity.

Oct. Thurs. 28; Lost 28 head of cattle last night on account of the prairie fire. Some of the hands are hunting for them this morning while the rest drive the remainder of the herd to town, to ship as they have been sold. All hands have been paid up and discharged. Only 3 head of the lost cattle have been found. So I have lost 25 head here, besides several on the way. I also had to kill some for beef on the way. So my sales amount to $4,664.00. Leaving me, after paying my hands and other expenses $4,068.00 currency. Camp 51.

Nov. Fri. 5; About 9 A.M. the train, consisting of 12 wagons, and 19 loose horses with 2 hands to drive them started from the old camp on Turkey creek, 5 miles from Abilene, to our home in Texas.

As well as a staging point for the Chisholm Trail, Palo Pinto can claim to be the origin of the Goodnight-Loving Trail, which headed west in a roundabout way toward markets in New Mexico and Colorado. During the Civil War the United States government forcibly removed Apache and Navajo Indians from their homelands to Bosque Redondo, near Fort Sumner on the Pecos River in New Mexico; after the war the agency became a profitable market for Texas beef. Even more lucrative were mining camps and booming towns in Colorado, where Oliver Loving had taken a herd before the war. Loving and his young friend Charles Goodnight, both Palo Pinto men, put together a herd in 1866 and threw them on the trail toward New Mexico with historic results.

Two thousand cattle transporting themselves to market require many things beyond the trail boss and his ten or twenty cowboys—grass and water first and foremost. Fort Sumner lay northwest of Palo Pinto 350 miles, but between point A and point B the Llano Estacado stood as an arid barrier. Buffalo grazed over its open grasslands and could find water here and there, and Kwahadi Comanches had made it their austere home—the tribe was also known as Back Shade, for turning their faces from the sun on the treeless plain, providing shade with their own backs—but few Anglos had dared venture over it. A caravan of Texans attempted to establish a trade route from San Antonio to Santa Fe during the days of the republic, and their mission was a total disaster. California gold seekers were accompanied by Colonel Randolph Marcy on an 1849 expedition across the northern Llano; Marcy, searching for a route west suitable for emigrants and a railroad, returned along its desolate southern perimeter. The interior of the Llano Estacado, however, was all but impassable, and Marcy's routes saw limited traffic.

The most practical trail from Palo Pinto to New Mexico and Colorado followed the Butterfield Overland Mail route, established in 1858 to carry mail between St. Louis and San Francisco. Butterfield stages rolled southwesterly across Texas, swinging above Palo Pinto along the line of frontier forts, down to the Concho River, and from there across to the Pecos—a trail long followed by Comanche raiders headed down to Mexico. New York *Herald* correspondent Waterman L. Ormsby, the only passenger to make the first historic trip, described the stage across the lower plains to the Pecos River: "To the emigrant this fearful journey is fraught with many terrors. He sees in his imagination his cattle, and perhaps himself, suffering all the inexpressible pangs of thirst-pains, more unendurable than flaming fire or bleeding wounds. . . . For a distance of seventy-five miles—the width of the plain at our crossing place—not a drop of water could be procured for all the wealth of the world." Despite such dangers, Overland Mail stages operated successfully until 1861, when the Civil War ended their service; their routes were well known to frontier cattlemen.

"It is generally believed that Goodnight and Loving were the first men to drive a herd up the Pecos in 1866. But that is a mistake," recalled W. C. Cochran. "The Wylie brothers—Bob, Henry, and Tom—and Captain John Henderson drove the first herd of one thousand steers that ever went up the Pecos in the fall of 1865." Cowmen from southern Palo Pinto County contributed cattle to this pioneering herd, and Cowden neighbor Irbin Bell went along: "We went a little south of west and struck the Concho River. We came up the Concho and crossed the plains to the Horsehead. Bob Wiley, an old-timer at Palo Pinto, bossed the herd. We came through Castle Gap on the way to Horsehead Crossing, having a ninety-mile dry drive. We lost no cattle but drove pretty near all the time on that stretch. We turned up the Pecos River from there to Fort Sumner. . . . We started in October on our drive out and were gone about four months, getting back in February. The biggest part of the herd was put up in Erath County and right on the line of Palo Pinto and Stevens. Stampedes did not bother us much. We thought we were making lots of money, as they paid us sixty dollars a month. We saw no sign of a wagon or anything up the Pecos. Ours was the first herd, and Bob Wiley led the way." Bell's matter-of-fact description suggests the frontier attitude toward what was elsewhere—in newspapers and dime novels back East—gripping adventure that prompted countless young men to head west to herd cattle.

The next June, Oliver Loving and Charlie Goodnight started a herd of two thousand head over the same trail. "Travel by the Emigrant Trail to California had removed some doubts concerning the country they faced," J. Evetts Haley wrote in *Charles Goodnight*, "but none of its difficulties and dangers. They trailed out into a tried, but still an uncertain, land." Goodnight later referred to the Pecos country as "the graveyard of the cowman's hopes," not only for the losses of cattle sustained on the trail but also for the death the next year of his partner, killed by Comanches. Their first seven-hundred-mile drive was dramatic enough—skittish Longhorn cattle, difficult terrain, threat of Indians—but they successfully sold their steers to government agents in Fort Sumner; Goodnight put together a second drive later that year, helping establish himself as the man who defined the western cattle trail. Other trail drivers from Palo Pinto and the Cross Timbers regularly took herds out to the Pecos and up to Colorado. "In the spring of 1870, Irb Bell and Tom Stuart started from Captain Frank Cowden's ranch on Rock Creek for Colorado with a herd of eight hundred cattle," recalled W. C. Cochran. Bell remembered the drive with slightly different details: "In 1870, me and Mr. Cowden made up a herd. We bought some and had fourteen or fifteen hundred head ready to go by August 1. Our horses and cattle were ready to go when the Indians came in and stole our horses and we had to go down by Weatherford and buy some. We were detained a week. We were in the south part of Palo Pinto County on North Fork Creek when this happened." The herd eventually sold in Fort Sumner, New Mexico.

So the Cowdens and other Palo Pinto cattlemen could point their herds either west on the Goodnight-Loving Trail or north along the Chisholm Trail. In twenty years, more than five million head of Texas Longhorns passed up the trails to Kansas cow towns; figures for the Goodnight-Loving Trail west are much lower, though still substantial. Many cattle went to slaughter, and others went to stock new ranches up in Wyoming and Montana. There's no doubt that new markets open to Texas cattlemen after the Civil War, combined with abundant supplies, created explosive growth in the cattle industry and great opportunities for the individuals involved. All the Palo Pinto Cowdens—William Hamby and George Franklin, their sons and daughters and sons-in-law—were building futures in the cattle kingdom, which would help bring disaster to the Plains Indians.

In *The Frontier in American History*, Frederick Jackson Turner took the retrospective view of events in the West in the last half of the nineteenth century: "Watch the procession of civilization, marching single file—the buffalo following the trail to the salt springs, the Indian, the fur-trader and hunter, the cattle-raiser, the pioneer farmer—and the frontier has passed by." White men and cattle were "destined" to replace Indians and buffalo on the Great Plains, as private citizens, business, and government created an irresistible force sweeping west. Federal military resources— officers and troops, modern weapons, hostility—shifted from Civil War battlefields to western territories in support of prospectors, ranchers, and settlers. Railroads pushed steadily west—the transcontinental railroad was completed in 1869, and other tracks were laid to transport western beef— and brought buffalo hunters who destroyed the great northern and southern plains herds. In *The Extermination of the American Bison*, William Hornaday related: "During the great slaughter they each killed from twenty-five hundred to three thousand buffaloes every year. With thousands of hunters on the range, and such possibilities of slaughter before each, it is, after all, no wonder that an average of nearly a million and a quarter of buffaloes fell each year during that bloody period." Considering all hunting, experts estimate that from 1870 to 1874 a stunning 2.5 million buffalo were slaughtered each year for hides, tongues, bloodlust, and the elimination of the fundamental basis of Plains Indian culture. When the Texas legislature met to consider laws to protect the southern herd, army general Philip Sheridan objected, "These hunters have done more to settle the Indian question in the last two years than the entire U.S. Army in thirty. They are destroying the Indians' commissary. Let them kill, skin, and sell until the buffalo is exterminated. Then your prairies can be covered with speckled cattle and the cowboy who follows the hunter." The slaughter went on. Hornaday wrote: "By the close of the hunting season of 1875 the great southern herd had ceased to exist. As a body, it had been utterly annihilated. The main body of the survivors, numbering about ten thousand head, fled southwest, and dispersed through that great tract of wild, desolate, and inhospitable country stretching southward . . . [across] the Llano Estacado, or Staked Plain, to the Pecos River."

Whatever grand historical forces were at work in favor of American pioneers, individual families like the Cowdens still faced an uncertain frontier. Isolated raids continued to plague Palo Pinto as desperate, vengeful Comanche and Kiowa braves slipped down the Brazos from camps on the southern plains and off reservations in Indian Territory. In 1869, Shepley Carter, William Crow, and John Lemley were killed; Marcus Dalton, James Redfield, and James McCaster died in 1870; up in Young County in 1871, seven teamsters were killed in the Salt Creek Massacre; in 1872, Chesley Dobbs was found outside Palo Pinto "killed, scalped and stripped of his clothes"; Jessie Veal went fishing on the Brazos in February 1873, stumbled upon Comanche raiders, and was killed at the mouth of Ioni Creek—one of the last Palo Pinto residents to die at the hands of Indians.

Federal troops and policy became increasingly aggressive and successful, pursuing hostile bands of Indians deep into formerly safe territory, out where the Llano Estacado broke into a series of rough, protected canyons at the headwaters of the Brazos and Red rivers. In 1874, Kwahadi chief Quanah Parker's Comanches were repulsed by buffalo hunters at Adobe Walls in the Texas Panhandle. Hunters had moved onto the Llano the previous year, after slaughtering out the buffalo in Kansas; their arrival raised the stakes for the Kwahadi and remnants of other bands still resisting white encroachment. When superior firepower turned back the Comanches—another reputed

"stand" by whites, but in truth the last stand of southern Plains Indians—
the Texas Indian wars were all but over. That fall, U.S. troops, especially the
Fourth Cavalry under Colonel Ranald Mackenzie, followed up with a
decisive campaign at Palo Duro Canyon, where renegade Comanches were
gathered. Mackenzie drove the Indians off and, more importantly, captured
and then killed their entire horse herd, fifteen hundred animals, destroying
both the transport and spirit of the Indians. By the next June, in 1875, the
last stragglers had joined other Comanches on the reservation in Indian
Territory. The tribe—estimated at twenty thousand persons a hundred
years before, but probably no more than ten thousand when the Cowdens
moved to the frontier—by 1875 numbered under two thousand souls.
The Comanchería was forcibly abandoned by its people.

The year that my great-grandfather Eugene Pelham Cowden was born—
1875—the twelfth and last child of William Hamby and Carrie Cowden,
the southern Great Plains were swept clean of buffalo and Indians. The
days of cowboys-and-Indians had passed, which is to say that the day of the
Plains Indian had passed. Even though Sitting Bull and Crazy Horse and
Gall would wipe out the Seventh Cavalry at the Little Bighorn the next
year, and other battles remained to be won and lost, the Plains Indians'
defense of their country was doomed. On the Comanchería, advantages
that Comanches had gained from their long, valiant adaptation to their
environment were gone, and their long, valiant resistance to white encroachment upon their territory was over.

Yet could the demise of one party occur without the other? Both Indians
and cowboys depended upon their herds, buffalo or beef cow, and the
wide-open grasslands that sustained them—could one realistically expect
to flourish where the other failed? Did not both need conditions that
history was unlikely to provide? Would not civilization destroy both? If
Turner's model of an evolving, disappearing frontier was accurate, open-
range cattlemen would follow Indians following buffalo into decline, if
not complete extinction.

However impossible to imagine in 1875, the birth of Gene Cowden
was end as well as beginning—he would ride the coattails of Comanches
into the sunset. He would join his older brothers in creating the great JAL
Ranch on the free ranges of New Mexico and West Texas, and he would
see its dissolution as the frontier shrunk like a waterhole on the Llano
Estacado in summer, margins receding day by day. Ahead was a blaze of
glory for the Cowdens, and they remain today a successful family, but
their way of life seems like a heroic battle for the past rather than the

future. Like the Comanche warrior who removes his moccasins as a signal to friends and enemies that he will fight to the bitter end where he now stands, so stand the Cowdens on their ranches in the Southwest. Deep inside themselves how can they doubt that cowboys and Indians, who share the same world, share the same fate?

JAL outfit mounted, Pecos River country, Texas, 1895. Courtesy of Haley Memorial Library and History Center, Midland, Texas.

Cowden outfit mounted, Cowden Ranch, New Mexico, May 2000. *Left to right:* Sam Cowden, Juan Estrada, Bill Mitchell, Abby Cowden, Hannah Cowden, Guy Cowden, Jerry Young, Pete Marez, Mark Wheeler, Souli Shanklin, Seth Shanklin, Herman Martinez, Ty Monisette, and Payson Marez. Photo by author.

JAL cowboys, Pecos River country, Texas, 1895. Cattle were gathered and held on open range, where calves were dragged to the branding fire, fueled by mesquite wood, and branded to their mothers' brands. General roundups brought together branding crews, or "reps," from different outfits. Courtesy of Haley Memorial Library.

Cowden Ranch branding, May 2000. *Left to right:* Mark Wheeler, Sam Cowden, and Ty Monisette, preparing to drag calves in the branding pen. Visiting cowboys have the honor of dragging calves. Photo by author.

Souli Shanklin dragging calf to be branded, Cowden Ranch, May 2000. Calves are first stripped from their mothers, since most calves are branded with the Lazy 6; Herman Martinez's calves are branded HM. Jerry Young in background. Photo by author.

Flanking a calf, Cowden Ranch, May 2004. Jake McCorkle drags calf to flankers Pete Marez *(left)* and Jake Shanklin *(center)*. One man "tails" the calf, the other grabs the rope, and they pull in opposite directions to drop the calf. Photo by author.

Working a calf, Cowden Ranch, May 2004. *Left to right:* Sam Cowden, Michael Pettit branding, Abby Cowden vaccinating, Pete Marez and Jake Shanklin flanking, Hannah Cowden vaccinating, and Jake McCorkle dragging. Each calf can be worked in less than one minute. Photo by Kathy Cowden.

Leonard Lujan branding, Juan Estrada flanking, Cowden Ranch, May 2000. The Lazy 6 brand goes on the left cheek and left hip of cattle. Photo by author.

Castrating a calf, Cowden Ranch, May 2000. *Left to right:* Payson Marez, Kasey White flanking, Bill Mitchell castrating, Ty Monisette Flanking. Payson will apply wound spray after castration. Photo by author.

JAL cowboys castrating, Pecos River country, Texas, 1895. Courtesy of Haley Memorial Library.

JAL outfit camped, Pecos River country, Texas, 1895. Courtesy of Haley Memorial Library.

JAL chuck wagon, Pecos River country, Texas, 1895. *Lower right:* Eugene Pelham Cowden sleeping, Spence Jowell reading, and Rorie Cowden behind Jowell. Cowboys are gathered in the shade of the chuck wagon; the plains were treeless. Beef hanging on wagon wheel would be wrapped and stored in chuck wagon bed to avoid spoiling. Courtesy of Haley Memorial Library.

Taking a break, Cowden branding pens, May 2000. *Left to right:* Mark Wheeler, Ty Monisette, Souli Shanklin, Bill Mitchell, Pete Marez, Herman Martinez, and Juan Estrada. Photo by author.

Another break, Cowden branding pens, May 2001. Geronimo Armendadez *(left)* and Jake Shanklin. Weigh house is on left. Photo by author.

Tumbleweeds

January 2001

January 23, 2001—Back in the saddle again. I'm flying out to New Mexico for a little winter work, which is mostly feeding cattle, a considerable task when you've got eight hundred head scattered over fifty thousand acres.

My pony is a Delta Boeing 727—manufactured in Seattle?—where I am served a "snack," which includes a Coke from Atlanta, a Sara Lee sandwich, a bad apple, and a pack of Poore Brothers Desert Mesquite Bar-B-Que potato chips made in Goodyear, Arizona. The Desert Mesquite Bar-B-Que artwork features a saguaro cactus in the foreground, blue mountains under a yellow moon in the background, and a coyote chasing a roadrunner across a desert landscape. I note how everything comes together from everywhere these days. Little is local. What did the Cowdens have to eat, traveling from Alabama to Texas, Shelby to Palo Pinto, Palo Pinto to Midland? They would have lived off the land around them, cooking rabbit over a mesquite fire. As Guy Kellogg said as we were driving out to Palo Pinto from Dallas, "I eat what I kill."

That metaphor harkens back to the literal self-sufficiency of our not-so-distant ancestors. It wasn't that long ago, although the pace of modern life—faster than the speed of sound, aspiring to the speed of light, jamming more and more bits of data into smaller and smaller spaces—can produce amnesia or indifference about all but the most immediate events. The current dismissal of patient accuracy is "Whatever." I think of Mark Twain's comment: "The difference between the right word and the nearly right word is the same as that between lightning and the lightning bug." Lightning, lightning bug, whatever.

January 24, 2001—Santa Fe, trading post of the Southwest for almost four centuries. Best to remember that fact before lamenting the tourists wandering high-priced shops around the plaza, Indians selling goods from blankets under the *portal* of the Palace of the Governors, and the sprawling flea market out beyond the opera house—the whole shop-till-you-drop atmosphere here. I'm heading to the ranch tomorrow and taking supplies hard to come by in Santa Rosa, fresh fruit and vegetables mostly. In the old days, custom was to drop off something at ranches you passed when returning from town with supplies or from hunting with game. Everyone welcomed a letter or a little venison. I'll stalk the aisles of Albertsons for avocados, jalapeños, oranges, whatever.

Today I'm also looking for books and maps, particularly ones from the period of Spanish and Anglo exploration of the Southwest. I've been able to find many things back East, but I believe there's treasure here locally, and maps that will lead me to it.

Bedridden all afternoon, struck down with some bug that rode me from Massachusetts. Fever, chills, swoons that laid me out flat under the quilts. So I looked back at the diary of James Baker, who made a trail drive from Palo Pinto to Kansas with William Hamby Cowden. I think it was W. H. Senior, rather than his son. The drive was in 1869, when Senior was forty-three and Junior sixteen; either or both might have gone. "Wm. Cowden came very near drowning, and it was with difficulty he was rescued," wrote Baker of crossing the Red River. Three weeks later, "Wm. Cowden is chilling," which could be anything, given his dunking and general exposure on the trail. The next week, "Mr. Cowden who has been chilling is about well." Since he's "Mr. Cowden," I assume it was W. H. Senior with a cold, flu, pneumonia, whatever. Had he drowned in the river or expired on the trail, I myself would not be here chilling—my great-grandfather was not yet conceived. I'm glad Wm. Cowden was tough enough to survive and sire a few more sons, and from his grit I take courage.

In addition to Baker's diary, I've been looking through books I found this morning, including Josiah Gregg's *Commerce of the Prairies*. The map of his routes to and from Santa Fe in 1839 and 1840 indicates that Gregg likely crossed the Cowden Ranch coming and going. The natural passageway from the mountains to the plains was used by Hispanic *pastores* and *ciboleros*, sheepherders and buffalo hunters who traveled from Santa Fe or San Miguel or Las Vegas (New Mexico, not the new burg in Nevada) to the Llano Estacado; their cart paths became routes for Anglo traders and travelers. After Gregg made journeys from Missouri and Arkansas—on the original Santa Fe Trail and the shorter, southern route along the Canadian River—he published *The Commerce of the Prairies, or The Journal of a Santa Fe Trader, during Eight Expeditions across the Great Western Prairies, and a Residence of Nearly Nine Years in Northern Mexico; Illustrated with Maps and Engravings.* A success, with editions published in New York, England, France, and Germany, Gregg's book appealed to the broad and deep interest in the American West, with its literal and figurative possibilities—adventure, novelty, and profit.

A subsequent, more practical work was Captain Randolph Marcy's *The Prairie Traveler: A Hand-book for Overland Expeditions; With Maps, Illustrations, and Itineraries of the Principal Routes between the Mississippi and the Pacific; Published by Authority of the War Department.* Marcy was ordered in 1849 to survey a route from Fort Smith, Arkansas, to Santa Fe, and to provide escort for a large party of gold rush emigrants. According to Grant Foreman's *Marcy and the Gold Seekers*—more treasure I found this morning—the Fort Smith and California Emigrating Company organized the expedition: "There were 479 in the company, traveling with seventy-five wagons . . . drawn by 500 oxen and as many horses and mules, besides hundreds of saddle- and pack-horses and mules. The caravan when united extended its length more than three miles along the road." Most of those private citizens have slipped unheralded into history, while Marcy became noted for blazing trails, although there were no trees on the prairies to blaze. Here in Santa Fe a fort was named for him, and a street downtown.

Marcy followed Gregg's route along the Canadian River, which east of the Cowden Ranch makes a sharp turn. To reach Santa Fe, Gregg and Marcy and others kept heading west, crossing from the Canadian to the Pecos, which they usually reached at Anton Chico. This route was in place even before the pastores and ciboleros arrived, before the very first Spanish explorations, as Plains and Pueblo Indians traveled to exchange various staples, mostly buffalo for corn. Indians, Spanish, Anglos—the three cultures of New Mexico—all intersecting on the Cowden Ranch: it seems remote and empty but was actually a crossroads of sorts. Perhaps there will be enough snow to highlight the cart tracks worn into Bogie pasture centuries ago. Sam showed me evidence of the trail last year but didn't know how to account for it. Now maybe I can answer his questions.

So tomorrow I'll join the procession of travelers—hunters and traders, herders and surveyors and emigrants—who passed through the gap where the Cowdens now hang tight, raising cattle.

January 25, 2001—Cowden Ranch. I'm still chilling, but in a comfortable bed in the bunkhouse. Which reinforces my admiration for old-time cowboys on the trail or range, exposed to the elements whether healthy or hurting, riding hard all day, sleeping in soogans—cowboy bedrolls made of quilts laid upon a canvas tarp, in which our heroes wrapped themselves like burritos. If it was rainy or cold, they folded the ends and sides over and completely enclosed themselves. Imagine how rank those bedrolls got after two months on the trail without showers, eating beans and beef, perhaps chilling or feverish. As Sam says, "Sleeping on hard ground night after night—ain't cowboy life grand!"

I drove over this afternoon and met Sam in Santa Rosa, at Jean's house. The house is on the road to Puerto de Luna, just before you cross El Rito Creek and get to the rodeo grounds. Located near the first hacienda established back in 1865, it's a sweet spot with big cottonwoods, a pasture beside the creek, a small set of corrals for the horses. Jean gets out to the ranch occasionally, but more often Sam or Kathy and the kids drive into town for supplies and services. At various times the kids have piano lessons, library programs, 4-H groups. Santa Rosa is a small town—population about 2,500—but doing reasonably well since I-40 has three exchanges there; motels, gas stations, and restaurants get pass-through business. It's also the county seat, so the courthouse is active and the center of town hasn't dried up and blown away. There are a handful of churches, a medical clinic, a couple of feed stores, a lumberyard, one grocery store. Just outside of town a new, privately owned, six-hundred-bed, medium-security prison burns its lights all night, a beacon of progress. Guadalupe County Correctional Facility has experienced one riot and two murders to date and is not universally popular in town, despite the jobs it offers.

Most remarkably for a town in semiarid country, Santa Rosa sports a number of spring-fed lakes, including Blue Hole, "Scuba Diving Capital of the Southwest!" Situated in a geological area called the Santa Rosa Sink, a large stratum of limestone that has collapsed from exposure to surface and subterranean water, Santa Rosa features more than a dozen smaller sinkholes—thus the motto "City of Natural Lakes." Blue Hole is an eighty-foot-deep artesian spring producing three thousand gallons per minute of cold, perfectly clear water. Only eighty feet in diameter, Blue Hole is nevertheless a big deal, its parking lot jammed in summer with vans from dive shops, instructors and students slapping around in scuba gear as if this were a Caribbean resort. They say you can dive to the bottom of Blue Hole at night and look up through eighty feet of water at the stars.

Along with Blue Hole, Santa Rosa's got Park Lake, Twin Lakes, Power Dam, and Perch Lake, which offers a twin-engine airplane submerged fifty feet deep, another playground for scuba divers. Also Tres Lagunas, Hidden Lake, Bass Lake, Post Lake, Rock Lake, and Swan Lake—a natural or artistic allusion? Lots and lots of water percolates up through the limestone. North of town is Santa Rosa Lake State Park, a big reservoir on the Pecos River completed in 1981. Lucky for me, that project called for extensive government research, which helps answer almost any question I might ask about the archeology, anthropology, or geology of this area.

Jean and I visited some before Sam arrived. After Rooster died, Jean took responsibility for the cemetery where her husband and son are buried, and she's involved with the church. Occasionally she heads down to Ruidoso, a mountain resort where the Cowdens bought a house years ago, but most of her time is devoted to her family, here and over in Texas. Last summer Patty and Souli gave her a Jack Russell terrier; Skip and Jean are now inseparable. Skip is a riot and Jean is unflappable. She's been in Santa Rosa or on the ranch since 1943, when her family moved here from Iowa.

Sam arrived to pick me up, pulling a flatbed trailer behind his pickup. "You better leave your car here. The road's in awful shape. You might get out there," Sam told me, "and never get back!"

We went first to check on Sam's road grader, which was leaking hydraulic fluid and had to be repaired. It's hard to get work done promptly. "Mañana," said Sam, who shows remarkable patience and a healthy sense of humor. There's little alternative. Tucumcari, 60 miles east, offers no more services than Santa Rosa; Albuquerque is 120 miles west, and Portales about the same distance southeast. "We're lucky to have anything here," says Sam. We went by another shop, owned by a high school friend of Sam's, Glen Gonzales, who installs and repairs windmills, among other mechanical work. "This is my *primo*," Sam told Glen when he introduced me. "He's writing a book about the ranch. Going to make us all famous. I told him the movie can star me and Shania Twain." Glen laughed. "But Kathy says it should star her and Brad Pitt." Sam picked up hydraulic fluid for the grader and arranged for Glen to bring his forklift to Pete and Leigh Ann Marez's place, where a truck was delivering tubs of protein supplement.

The supplement is cooked molasses and artificial protein—urea, also used as a fertilizer—that the cattle can lick when Sam and Herman can't get other feed out to them. They set out the 250-pound tubs at watering holes in each pasture, and the cattle get extra protein whenever snow gets deep enough to cover the old grass. "I'd say my cows have lost seventy-five pounds in the last month," Sam says. That comes at the worst possible time, since they are due to start calving the first of February. "Over at the Bar Y, they had to pull a bunch of calves and shoot two or three cows that got down calving." Cows with difficult deliveries, often heifers having first calves, can experience numbed, "prolapsed" nerves in their hips and can't rise. Unlike horses, which rise first on their front legs, cows get up ass first, and if they can't get their butts up, they stay down, leading to further numbness, weakening, and eventual death. The only solution is for a cowhand to grab the cow's tail, haul her ass end up, and stagger around with her after she rises. With luck and attention, feeling returns to the nerves. I've brought back prolapsed cows that way, but it was in Mississippi, where winters are milder and pastures much smaller. It's not always successful or easy in the best of circumstances. Sam has a bad right ankle that came when he and Herman were wallowing around behind a prolapsed cow, and Herman stepped on Sam's ankle, trashing the ligaments. Sam still wears a brace on occasion to support it.

So it's important to take your cows into calving season in the best possible shape. Better for the cow, the calf, the cattleman. But when snowstorms blow through the country every few days, as they have this winter, it's a struggle to keep your stock healthy. It's been hard to get feed out to the ranch on roads that are passable when frozen, but impossible when snow-covered and thawing. "Trucks have to make their runs first thing in the morning," Sam says, "while the road's still frozen hard, or not at all. That's why I have to pick up this feed and a drum of gasoline—nobody's been able to make deliveries to the ranch."

We went to Pete and Leigh Ann's to wait for the feed truck. We visited until the truck arrived, then Glen with his Bobcat. They got the protein tubs transferred to Pete's barn and Sam's trailer, and after a beer in the barn with Pete and Glen, we took off for the ranch, making the drive in about an hour. I'm glad to be here again, eager to see the winter routine. And routine is the major part of ranching life, however much the public thinks of it as adventure.

January 26, 2001—Done with lunch, time for siesta. My grandparents took a siesta every day of their lives, after "dinner" at noon. Paul Harvey's radio show would come on and Rosa would announce, "Dinner's ready." Their cook for years, Rosa was a black woman with thick glasses and a quiet, sure manner in her kitchen . . . and it was *hers*. She served breakfast before seven, then turned her attentions to dinner: roast or a chicken, vegetables, bread, pie, all from scratch. Did that big midday meal echo trail-driving days, when cooks served breakfast before dawn, packed the chuck wagon, drove to the next campsite, and prepared dinner? Country people in many regions take their big meal midday; siestas are more common in the Southwest, as in Mexico and Spain. Too hot to work, so take a nap and let the weather moderate. At least the Cowdens always did. Still do. "Cattle just don't work well after noon," Sam says. "I've tried it and I'm always sorry. They won't drive, they get hot and bothered, nothing goes right." So the cattleman follows his cattle's lead and shuts down midday. Dinner and maybe a quick nap. Then chores around headquarters, but rarely any cattle work.

Sam follows the customs of his predecessors but is also a man of his own time . . . sort of. He thinks about health in modern ways . . . sort of. To lose some weight he started on a high-protein, low-carbohydrate diet. "What could be better?" he asks gleefully. "Lose weight—eat meat!" A devoted carnivore, as you'd expect of a fifth-generation rancher, Sam has the Atkins diet to back up his natural affinities. Kathy and Jean both were skeptical about the health benefits: how could a diet of bacon and beef be good for you? But Sam eats vegetables and fruit, just very few carbohydrates. Our grandfather ate everything, all of Rosa's cooking—beef, vegetables, biscuits, dessert—and lived ninety years. "It was carbs that made Big Guy gain weight," says Sam. "All the protein, the steak and eggs, cholesterol— I think a lot of that's in the genes. So bring on the brisket!"

Sam is the picture of health, whatever the reason. I don't overlook the fact that he is physically active and infrequently exposed to contagious diseases. Pilgrims have traveled to the Southwest for their health for centuries. Josiah Gregg came to the plains to improve his health; Teddy Roosevelt came west (and recruited his Rough Riders from this area) to make himself "bully." And I found the following in Randolph Marcy's *Thirty Years of Army Life on the Border:* "Perhaps, no part of the habitable globe is more favorable to health and the continuation of human existence than this. Free from marshes, stagnant water, great bodies of timbers, and all other sources of poisonous malaria, and open to every wind that blows, this immense grassy expanse is purged from impurities of every kind, and the air imparts a force and vigor to the body and mind which repays the occupant in a great measure for his deprivations."

Look at me: fresh from the poisonous East and puny, chilling, stretching my siesta in the bunkhouse as long as I can. I've come to this immense grassy expanse to purge myself of impurities, to bring force and vigor to my body. Bring on the breezes, the brisket, the nap!

This morning I rode with Sam while he fed cattle. He tries to feed each pasture every other day, weather permitting. He takes one area, and Herman another; today Herman fed yearling steers and heifers up on the mesa, and we fed cows on the south side. The ranch has three feed trucks—three-quarter-ton flatbed pickups with automated feeders on the back—one for Sam, Herman, and whoever else is on the payroll. They're used for everything: feeding, hauling horses or cattle, checking water around the ranch. Sam and Herman both have their own pickups as well; Kathy has a four-wheel-drive Suburban she takes into town. I count six vehicles on the ranch, plus three stock trailers, a road grader, a tractor with a dump-bucket, an ATV. Parked in the big barn are two campers—one Sam's, one Herman's. Over at Alamito the old Power Wagon, a four-wheel-drive pickup from ancient days, sits in the shed like some green Buddha of Go, although it never budges.

This is where I first learned to drive, sitting beside my grandfather on a ranch road far from any possible collision. In the distance were blue mountains with snowy peaks; at my feet were the accelerator, brake, and clutch. I peered over the top of the steering wheel and tried to get the pickup going. I bucked and stalled it, bucked and stalled it. The mountains sat there, waiting. When at last we lurched into motion, Big Guy would tell me, "Keep it in the ruts." Cactus passed slowly beside me, the mountains hardly moved. "Give it gas," said Big Guy. "Keep it in the ruts." I was ten years old and sensed some impossible contradiction in his instructions.

Sam drove this morning, of course, which meant I'd have to open gates, as in the old days. Except those days are gone and Sam has cattle guards at fences on his main roads. You don't have to stop, open the gate, drive through, close the gate, drive on. "Jean gives me a hard time about these cattle guards," says Sam, "but she wishes she had them back then." Back then, Jean told me yesterday, she often helped Rooster feed. "You know we didn't have overhead bins and all that," she said. "Lots of times it was Rooster in the front of the pickup and me in the back, pouring out feed. Or hay, pitching off hay, or throwing off salt blocks or whatever."

Nowadays, Sam and Herman pull their trucks under one of the bins and load up in under a minute. "The truck, every three weeks or so, if it can get here, brings out twenty-four tons," Sam says. Friona Industries in Lubbock manufactures the Hi-Pro feeds the Cowdens have used for years, in various levels of crude protein, fat, fiber, and minerals—cottonseed or soybean meal mixed with goodies like molasses, calcium, phosphorus, potassium, vitamin A, salt. "We're feeding twenty-percent-protein range cubes, eight pounds per head, three times a week. That varies according to conditions. We usually start in December at five pounds before they begin calving, then get up to ten pounds when we turn the bulls in. Don't stop until we brand in the spring."

Sam and I went first to Shipping pasture, where his replacement heifers are due to have their first calves. He calves out about a hundred heifers each year and keeps them close by, since they experience more calving problems than mature cows. "We can't watch them all the time, and we've tried different bulls—Angus, Longhorns, and now Corrientes—looking for a smaller, easier calf at birth. Buyers didn't like the Longhorn crosses. The calves came out speckled, skunk-tailed, everything. Maybe these Corriente bulls are the answer. Most of the calves are black, so we don't get penalized on the price." Descendants of Spanish cattle brought to the Americas in the fifteenth century, Corrientes are cousins to the Texas Longhorn and to Mexican criollos, mixed-blood cattle noted for their varied color. Corrientes are small, horned cattle now used frequently as roping and bulldogging stock in rodeos; Sam bought a few head two years ago because he was interested in team roping and knew there was a healthy demand from rodeo producers for calves. "Then too, I had some country over beyond Alamito—Mesa pasture—that was rough and hard to get to, winter, summer, anytime. I thought maybe we could run Corrientes in there, not fiddle with them but once or twice a year, see if it worked out. John Brittingham at Park Springs is running half-Longhorn cattle and doesn't feed them at all, doesn't touch them until he weans his calves." Corrientes, true to their Spanish/Longhorn heritage, are noted for their hardy, self-sufficient character. "It seems like more and more we're having to vaccinate for disease, pull calves, guard against predators—all sorts of things we haven't always done. Corrientes can take care of themselves. I'm not trying to make a lot of money off them, just see whether they can fit in here somehow."

After feeding the heifers, we moved on to Bogie pasture, where Sam has some black baldies that now form the heart of his herd. After barbed wire and windmills made it possible to control and manage pastures, ranchers imported British beef breeds to improve their herds. Texas Longhorns had adapted well to the open ranges of the West but were slow to mature and did not "grade"—produce a carcass of quality beef—particularly high. "They talk about this lean beef—'Everybody wants lean beef.' They *don't* want lean beef," Sam claims. "You read this every day—lean, lean, lean—but those carcasses don't bring as much money. They don't sell as good. You don't want a tough steak. Texas Longhorns were famous for being tough." Crossbreeding almost wiped out the Longhorn strain, which has seen a resurgence that is part nostalgia and part economics; Longhorns remain fertile, easy-calving, hardy, long-lived cattle that have survived on southwestern ranges for four centuries. British breeds, however, are the favored beef cattle, since they grade better and mature much earlier. You can put an Angus steak on the table in two years; Longhorns traditionally took four years from birth until slaughter.

Guy and Rooster Cowden were Hereford men, like many early ranchers in Texas and the Southwest, but Sam finds black cattle "hard to beat." He says, "They don't have eye and bag problems like Herefords. The cows are easy calvers, good milkers, attentive mothers. Buyers love black white-face calves because they're going to finish quick and grade well." Both consumers and producers influence trends in the cattle business. Heavy, fat beef was once in demand, but consumers have moved toward leaner cuts, and feedlots want animals that don't take so long to finish—those preferences influence producers. Breeders have to consider factors like fertility, birth and weaning weights, hardiness, resistance to disease and pests, disposition, and longevity, as well as market forces based on conformation and grade. Brahman and Brahman-cross cattle do better in hot, buggy Gulf coastal areas than straight British breeds, but not so well on the plains or in feedlots. The various breeds all have their advocates, including national associations devoted to improving and promoting their products. "The Angus people have really done the job," says Sam. "Look around, you'll see 'Angus Beef' marketed more effectively than anything else. That works back through wholesalers and feedlots to buyers out here looking for cattle."

Sam's black and black whiteface cows were scattered across Bogie, in pockets where they could find some old grass. Sam can spot cattle in the far distance—Charles Goodnight called it "plains vision"—long before I do. "There they are," he said, and I saw only empty landscape. But Sam knows where to look, where cattle tend to collect, depending on conditions. We drove across the pasture toward them, Sam honked his horn, and they came running. Classic Pavlov. As he gave them all time to arrive, he stepped out of the pickup and walked through the herd, looking them over for possible problems—bad eyes or bags, poor condition. His presence also helps gentle them for other times when they must be worked. After he returned, he threw a switch in his pickup and drove along slowly, the automated feeder *thunk-thunk*-ing over and over, dispensing a preset dose of cake with each *thunk*. Cattle trotted behind the pickup, stopping to feed, others taking their place, stopping, until a long line of cows stretched out behind us, noses to the ground, gobbling range cubes. "You been counting?" Sam asked me. "That was about eighty clicks," he said, switching off the feeder. "Let's get a count on the cows."

Counting cows is something Sam and Herman do often, trying to get a sense whether they're all in place and healthy. We drove along the line of feeding cattle, each of us making a silent count, by ones or twos or fives, whatever was possible given the circumstances. "Seventy-three?" I asked. "Seventy-three," agreed Sam. "The others are probably down around Lucky tank."

Today, in addition to feeding range cubes, we set out the protein tubs Sam brought out yesterday. "They don't like this stuff near as much as cake," he said. "They will almost always come for cake. But maybe this will give them a little something extra, something when we can't get around to feed. I'm trying to keep them in good shape while they're bringing these calves. Once that calf's on the ground and nursing and we turn the bulls back in, it's hard on them. Nursing a calf and breeding back—that all requires a cow in good shape. Tough to do at the end of winter. But if she doesn't deliver a live calf or doesn't breed back, well, that cow isn't earning her keep. I have to take care of them so they can take care of us."

All around us are the open, mild faces of Sam's cattle, watching him, chewing, their breath warm and sweet on the cold air. In university classrooms, they call this Animal Husbandry.

January 27, 2001—Snowing here, snowed all day, may snow all night and tomorrow. Not at a terrible rate, but it keeps piling up. There's low pressure full of moisture to the east, a high to the west sliding toward that low, and this storm brewing between the two. The ranch is covered with snow, not deep, perhaps six inches right now. In the days of open ranges, winter storms caused cattle to drift southward—spring roundups were held to retrieve them and work new calves—but now fences hold everything in place and ranchers have to help their stock battle the elements. It's not easy making the rounds to feed and break ice on water tanks, but Sam gives me the ranching perspective: "Once I was moaning and groaning to Big Guy about the snow, how much trouble it was, and Guy said, 'Sam, don't worry about the snows you get. Worry about the snows you don't get.' On a ranch out here—any moisture, anytime."

It's been a severe winter, or perhaps a normal one, following two or three dry winters as a result of El Niño. Sam says the manager of Singleton's Pecos division, who is from Nevada, has been griping about the snow, which he didn't have to deal with his first years down here. He thought he was out of snow country. Not. Sam remembers their worst winter as 1961, when they were living at Alamito, always colder because it's down in that big valley off the mesa. "It was thirty below," he says, "with thirty-six inches of snow on the ground. It got so bad after one storm that Rooster rode horseback all the way over to Park Springs to get help. I've always wondered why he did that. Left his wife and kids to ride over there, and it wasn't much closer to town than where he was. I guess he felt he had to do *something*." Alamito began as a line camp. It's way out there, almost forty miles to pavement, and nothing in sight from the house. A long, snowy winter would be difficult—cabin fever could really work on a cowboy used to outdoor activity.

Sam and I fed again this morning, and now he's gone to town to pick up Abby and bring back the road grader. I planned to go with him, but this flu still has me feverish and chilling, so Herman went instead. I rested, then talked with Kathy and the kids about homeschooling. It's no novelty that the Cowden kids are homeschooled; ranchers and other frontier folk have faced the problem of educating children for generations. Cowden children got perilous and spotty schooling in Palo Pinto during the Civil War; few of them studied past the age when they could begin cattle work in earnest, about ten or twelve. My great-grandfather went out to the JALs at that age, ending his formal education. Guy Cowden was shipped off to New Mexico Military Institute for school, and Olive Cowden told me last year how she left Valentine, Texas, for boarding school in El Paso: "I was eleven years old. There was just no choice back then. I stayed in El Paso—I'd go home and visit—until I was married."

My own mother never attended a conventional school until she entered high school, at age thirteen. Before then, she and Rooster went to school on the ranch, studying first with my grandmother, then under a succession of governesses who taught them and neighboring ranch children. "I went in for exams at the end of the year," she explained, "but that was all." Her younger sister, Mumzy, was the first child in the Cowden family, and mine the first full generation, to receive conventional schooling—startling when you consider that I began first grade in 1956.

Distance, rather than philosophical differences or failures of the school system, was responsible for the homeschooling of my mother and those other ranch kids. These days it's more likely that parents, rural or urban, will object to some element of conventional public or private education—lack of religious instruction, for example—and turn to homeschooling. "A lot of the curriculum guides you see at homeschool conventions have a Christian component," said Kathy, who goes every year to see what materials are available. "But there are tons of different approaches you can use."

Kathy taught school before she married Sam, just as my grandmother was a schoolteacher. One evening last May in the bunkhouse the cowboys talked about ranch women. "Most of those old ranchers started out batching," Souli said. "And who they married—when they found a sucker—were schoolteachers. My grandmothers on both sides were schoolteachers. My mother was a schoolteacher. Both of my aunts on the Shanklin side were schoolteachers. My great-aunt on the Shanklin side was a school-teacher. Patty, she's got her teaching certificate." Sam observed that Souli too had a teaching certificate—a shocker if you weren't aware of Souli's curiosity and avid reading. "Closet intellectual," I said. "Damn dark closet," Souli replied.

Kathy enjoys teaching: "I love it. I'll teach them just as long as they want to stay out here." Sam chose not to attend New Mexico Military Institute, but his sisters both went off to boarding school in El Paso, which remains an eventual option for the Cowden kids. The opportunities in Santa Rosa, where Sam went to high school—one of a handful of Anglo kids in an overwhelmingly Hispanic school district—are limited. There's no way to dodge the issue of race out here, just like in the public school system of New Orleans, where I grew up and where white students flock to private and parochial schools. Poor black or Hispanic school systems have trouble attracting and keeping good teachers and routinely lose more affluent white students to alternatives—no news there. Since Kathy is qualified and willing, the Cowden kids are homeschooled and happy with it. They are curious, active, well adjusted; my own kids, who attend public high school in Massachusetts, are always impressed when they visit the ranch. Homeschooling, like most teaching, like all good teaching, inevitably throws into relief the values of teachers, in particular disciplines and in general. What's important about language, math, history? What's important in a life? Kathy and Sam can't avoid those questions the way some parents can when they send their kids to school. I have little respect for politicians who tout "family values" for partisan reasons; I have real admiration for the Cowdens, who have built their lives around family. Cheap rhetoric doesn't work out here, where there's a premium on good behavior. The Cowden kids are not at all repressed but are extremely well-mannered, just as Cowden horses and Cowden cattle are well-mannered. I don't begin to apologize for grouping humans, horses, and cattle together—life on ranches doesn't let you distance yourself too far from your stock and trade. You'd *better* have horses that won't buck you off and cattle that won't run you over. And you'd better have solid values. Hard work, steady and patient attention, awareness of the past and future within the present, intelligence, flexibility, good humor, manners—the list grows as I spend more time out here, my own values called into consideration.

January 28, 2001—Snow showers linger today. A second wave of heavy snow slid by to the north, which means Pete, Leigh Ann, and their kids will make it out after all. We'll have company for Super Bowl Sunday. Baltimore versus New York. That might excite the eastern media but doesn't play big out here, which is Denver or Dallas country. Guy and Mushy were big Dallas Cowboy fans, but they were Texans. Rooster and Sam, especially Sam, developed loyalties to New Mexico, which maintains a rivalry with Texas dating back to the 1800s. When the Republic of Texas won independence from Mexico in 1836, it claimed everything to the Rio Grande. In 1841 an expedition of Texans left Austin for Santa Fe, hoping to establish a trade route and subvert Mexican authority there. The Texan Santa Fe Expedition almost perished crossing the Llano Estacado but managed to reach Laguna Colorado just east of here, where they were arrested by Mexican authorities. A number of Texans were shot or died from exposure when they were marched off to Mexico and imprisoned, further arousing Texan-Mexican animosity. When Texas was admitted as a state, it relinquished claims to eastern New Mexico, but the region is still pulled between the more Hispanic Rio Grande valley to the west and Anglo influences in Texas. Current New Mexicans, Hispanic and Anglo, often resent the money and/or manner of Texans past and present. I remember hearing on Santa Fe radio a song to the tune of the Chambers Brothers' "War," which went: "Texas—what is it good for? Absolutely nothin'." If God had meant Texans to ski, goes the joke in New Mexico, He would have made bullshit white.

This Sunday the Cowdens won't make the run into Santa Rosa for church, as they often do. Like formal schooling, formal religion was usually missing from life on ranches. Except for the occasional circuit rider, priests or preachers weren't available. Ranching families would wait for summer to gather with others to worship, when they'd attend "camp meetings" like the one down in the Davis Mountains of Texas. My grandparents spoke frequently of Bloys Camp Meeting, which confused me as a child—why would adults go to summer camp? Two years ago I went myself, in part to see relatives who might have information about Cowden history.

I was thoroughly surprised, first by the Davis Mountains, the "Alps of Texas," which lie across the Pecos River, southwest of the JAL Range. Fort Davis was established in 1854 to protect the frontier against Apache and Comanche raiders; the country is still rugged, beautiful, and remote. The

University of Texas's world-renowned McDonald Observatory is located in the mountains, far from any light pollution; the nearest major city is El Paso, 150 miles away. I drove first past ranches on the mesquite flats, up through oaks in the foothills where cattle and deer grazed, then past high pastures tucked below rocky peaks. Like most mountain regions in the Southwest, the Davis Mountains catch more rain than the surrounding semiarid plains. They were green and cool, and I could see why people flocked there for a retreat.

Bloys Camp Meeting, or Bloys Cowboy Campmeeting, began in 1890, founded by a Presbyterian circuit rider to bring ranchers together for services. Lots of the founding families are relations of mine, through blood and marriage. In fact, Campmeeting was often where young ranch men and women were first introduced to one another, where they found wives and husbands under the respectable umbrella of religion. My great-uncle and -aunt first met there, and my grandparents attended, although Guy Cowden was no Bible student. I figured if Big Guy went, anyone could.

I was shocked to come around a curve in the mountains and find before me countless little cabins scattered among the oaks, a host of cars parked across the highway. I arrived at noon, and people sat eating at picnic tables in large, open dining sheds, food cooking on barbecue pits and fragrant wood smoke drifting through the gathering. Folks wandered among the cabins and sheds, and children ran in packs here and there, the whole scene appearing more summer music festival than religious revival. But I found out there were four sermons daily at a big central meetinghouse—the tabernacle—and prayer meetings, Bible study classes, and children's events. Presbyterians, Baptists, Methodists, and Disciples of Christ were all represented. Whatever else went on, and no doubt it did, religion was front and center for the three thousand souls gathered there.

I stayed in Fort Davis, but Campmeeting members have their own simple cabins on the grounds—no phones, television, or radio, although they have electricity, running water, and plumbing. Utilities are provided by the association; nothing is bought or sold; all food is free. Area ranchers often contribute a beef to the meeting and bring cooks experienced in chuckwagon cooking on open fires. My cousin Jon Means helped run the Means-Evans cook shed, where almost everyone had some relationship to the Cowdens and some memory or story about the ranching heritage that was responsible for their presence now. Bill Cowden, grandson of "Uncle Billy," eldest of the Cowden boys who began the JALs, sat down with me to swap stories; George Newman gave me details about ranches in the

Midland country after the JALs dissolved and the Cowdens scattered. As at Rooster's funeral, I found myself in what felt like a time warp, life resurrected from the past—but for the pickups and cars parked across the road, it might have been 1890. The truth is, the warp wasn't in time so much as in space. I had traveled back to the physical ground of the Cowden story, to a landscape and its labors that had not changed as dramatically as elsewhere in urban, industrialized America at the beginning of the twenty-first century.

Now here I am again, lost in space, although we've caught up with time. Kathy has a television preacher providing religious instruction by satellite dish—the modern equivalent of a circuit rider. And later today we will see the Super Bowl from Tampa, have a big meal with stuffed jalapeños and brisket, and maybe spin a few windies.

January 29, 2001—Unable to return to Santa Fe today, as I planned. Sam was going to drive me to town after we fed this morning, but that became impossible when a "ground blizzard" swept across the ranch.

We headed out to feed first thing, trying to reach cattle we missed last week, in particular some heifers over at Alamito. The pastures and roads were covered with snow, not very deep, but the wind got up, blowing out of the west, then blowing hard, then *really* hard—thirty to forty miles per hour. It produced a surreal scene, windblown snow creating a whiteout three to four feet high, like a thick icy fog. The cab of the truck stuck up through the wind-driven snow like some forlorn island. I was nervous and thrilled and awed all at once. "I'd sure like to get something to these heifers. They haven't had anything in almost a week," said Sam, negotiating across the landscape more by memory than sight. He had to guess where the road was, then hunt for cattle, then find a spot where we could deliver their feed. "I kind of look for a windswept hillside, some bare ground, so we don't dump the cake in deep snow." Cattle aren't as adapted as their predecessors, buffalo, to snow country. With heavy, wooly coats around their heads and shoulders, buffalo are able to face the wind and root through snow after feed; cattle typically turn tail to the wind and have greater difficulty smelling and locating forage. If Sam and Herman weren't looking out for them, in a bad winter storm these cattle might pile up and die. I could see why, before fences were in place to stop them, great numbers of cattle drifted south in winter.

We found the heifers and a relatively bare spot to feed them, then drove into Eversaw pasture to feed the cows there. On the way, Sam lost the road and we got stuck. After trying to rock the truck back and forth in high, then low four-wheel-drive, Sam nodded, "We're stuck. Big time. I'm getting good at this." He got on his old but powerful mobile phone to call Herman, feeding up on the mesa. "I hope he's up there," Sam said. "Somewhere he can get reception. If not, *primo*, it is a long walk home." How far? "Probably six or seven miles from here. Pretty tough in these conditions." Actually, the wind had died a bit and we could see all around us—cholla rising from the snowy flats, junipers dark against the snowy mesa walls and ridgeline. Beautiful, but potentially deadly.

Sam was able to raise Herman on his phone. "Herman? We're stuck. . . . On the road from Alamito to Eversaw, just past Cap's Mill. I guess I should say '*off* the road.' I'm in the ditch. . . . *Bueno*." Sam hung up and told me, "He's coming. Good old Herman. Saving my ass a second time now."

Sam had plenty of time to tell me stories, like when Rooster made him walk home from Bogie pasture. "When we were kids riding horseback, Tommy used to torture me. Anytime Daddy wasn't looking, Tommy would come up behind my horse and whip his butt. You know what the horse would do—away we'd go! I'd be off, and Daddy would get mad at me for not controlling the horse. He didn't know Tommy was doing that. It was awful. After a while, Tommy had done it enough times so he'd just look at that horse—Macaroni—and there we'd go. And I'd get my butt chewed out. One day we were riding back from Bogie, and Tommy got Macaroni running and Daddy made me get off the horse, take my saddle off, and carry it back to the house. I was maybe ten years old. That crushed me, whenever I had to carry home my saddle that day." Only because I'm his cousin and can remember Tommy and him as little boys is it possible to think of Sam Cowden as crushed. He seems now a model of capable, sure action.

Then he told me about the time at Pruitt tank, almost as far as possible from headquarters, when Herman got out of his truck to check something. Trouble was, he didn't set the parking brake, and his pickup started rolling, slowly, then faster and faster, down the slope of the tank, out into the water. "Can't you see Herman, chasing that truck, trying to catch it before it rolled into the tank? We didn't have cell phones then. Herman had to walk all the way home. At least that was summer." I pictured the scene as if from above: Herman at the edge of a cattle tank where his pickup sat silently, hopelessly, out in the water. Nothing around him for miles on the ranch. Mountains rising off in one direction, plains stretching away in the other. One man and his lonely predicament.

Herman arrived to help us, Debbie with him in the pickup for company. He pulled us out of the ditch, full of good humor, since now he had something on Sam to balance his own misadventure. "Golly. Stuck the grader in the cattle guard, stuck the truck in the ditch," Herman said. "*El patrón* is on a roll, isn't he?" He wasn't going to let Sam off the hook, anymore than he'd leave him in a snowy ditch.

Sam and I continued up the mesa and found the Eversaw cows back in the junipers, where they had some protection. Many had green faces. "From eating bear grass," Sam told me. "You know it's rough when you see them like that. I don't know how they stand it, but they'll root down among those sharp spikes to get at the tender stalk. They'll eat yucca, cactus, almost anything if they have to. There's some protein in cactus, but also so much water they can't eat enough to do them any good." Black whiteface cows with green makeup stood around us, bawling for a meal while Sam waited for them all to arrive. If most people are comforted by their dogs, wagging their tails to greet any return, Sam has cattle to comfort him . . . and demand his attention: what's for dinner?

We headed into Bogie for our last handout and found the cattle just before the wind picked up again, blowing a gale now, buffeting the pickup, drifting snow, obliterating the ground in another whiteout. We took off for the house. As we approached the main road, we found Herman and Debbie broke down, truck hood raised as Herman tried to get the pickup running again. He was dressed in insulated overalls, a hood pulled over his head, only his face and hands exposed to the wind and windblown snow. His Fu Manchu mustache was coated in ice, like the beard of a cross-country skier. The wind was howling. Their son Ricardo was also there, with the pickup he'd driven out from headquarters. It too was stuck, slipped off the road into a drift. "Pretty big wreck here," Sam said. He cautioned me to open the door carefully, so the wind didn't catch it and rip it off its hinges, something Rooster always warned him against: never leave your pickup doors open, not in this country.

We couldn't get Herman's truck started again, but we did pull Ricardo's pickup out and everyone hustled back to headquarters to warm up and dry out. Herman told us later that windblown snow had soaked through his overalls, jeans, long johns, everything. "I was *wet*, man," he said, in the lovely lilting voice of Spanish New Mexico. More than anything he seemed amused. Because don't worry about the snow you get, as Big Guy said; worry about the snow you don't get. At least I'm not stuck out there in Eversaw or at Pruitt tank, wind shrieking, snow piling up around my vehicle, cattle bawling from hunger, cold numbing my body, my thoughts becoming more and more ethereal, music playing somewhere—harps maybe, coyotes howling.

January 30, 2001—Back in Santa Fe. This morning Sam drove me into Santa Rosa—no cakewalk. From years of grading, the road is lower than surrounding pastures, and drifting snow tends to collect in it. We had to stop at the front cattle guard, on an exposed ridge that often drifts badly. Snow that piled up yesterday had iced over, and the entire cattle guard was buried beneath a frozen drift. First we tried to open the road with shovels, but no go. Sam called Herman, who came with the grader to plow us out, then lead us down the road a few miles. "From the ranch to the Bar Y, the weather's often entirely different," Sam told me. "Drifts are worse, and the snow just never melts as quick. Once you get halfway in, things are always much better." I wonder how many times Sam has driven that road over the course of his life. Jean told me that when they lived at Alamito, she drove to the Bar Y and back—"Thirty-eight miles, twice a day, rain, shine, snow, whatever"—taking kids to the school bus. Consider the difference between an hour commute from the suburbs in traffic and an hour out here. What must the mind do in those different spheres—a jammed high-speed freeway, an empty ranch road?

When we got to town, I found out that the interstate had just reopened between Santa Rosa and Albuquerque. The highway from Clines Corner to Vaughn—just south of my turn to Santa Fe—was still closed because of drifting snow. I made it back, but the highway was especially deserted, highlighting the isolation of the West, with its tinge of danger and attendant excitement. And strange to be here in winter, after so many visits during summer. The country took on another dimension, its austere beauty magnified by snow cover that glowed blue in the distance everywhere—a cold mood I'd never seen before.

Now I've got time to record what Sam and Pete said about coyotes and trapping. Sam's too busy these days to trap, as he and Herman once did; bounties and the market for furs have also disappeared. Old traps hang on the fence behind the Bullpen, relics like bright lobster-pot buoys in a Cape Cod seaport. "Trapping's not easy," Sam says. "Think about it. You've got fifty thousand acres, and you're trying to get an animal to put his foot down on one square inch of that. Coyotes are leery sons of guns. Any human scent at all on those traps and you'll have nothing to show for your trouble."

The wiles of coyotes are legendary, of course, from crapping on traps to grinning taunts whenever a man encounters them without a gun. Indisputably resourceful and adaptable, they have extended their range over the years, as human incursions have extended into their territory. I think of coyotes as cousins to mesquite, another hardy species that's a scourge to most ranchers. Coyotes prey more upon sheep than upon cattle, more upon small game than upon any ranch stock, but they've earned the undying opposition of stockmen, who naturally enough don't want to lose even one lamb or weak calf to predators. Classic turf war.

Pete Marez works with county, state, and federal predator control programs, and his work takes many forms. Most striking are the occasional airplane or helicopter sorties to eliminate coyotes. Animal damage control programs pay half the cost of the plane; the rancher pays the other. Sam told me last spring they'd killed seventeen coyotes in a single morning's flight. "We didn't know until four in the morning we were getting the airplane. I try to book it a year in advance, always at calving time, but you never know because of demand and weather. You need good sunlight and no wind, and the day was perfect. It takes people on the ground along with the airplane, which has a pilot and a hunter with a shotgun. The men on the ground use a rabbit distress call and the trapper has a coyote call, a horn that imitates a coyote bark. Once you locate the coyotes on the ground, you radio the plane, and they go get them. We concentrated on areas where we have cattle calving. We started about 7:30 and by 10:00 had seventeen. Makes you wonder how many are out there, if you can kill seventeen in under three hours."

Was there any way to tell about his losses? "You can't tell exactly. Coyotes eat their kill right away, so unless you spot them, there's no way to know." Did they still hang dead coyotes on the fences, as Big Guy used to down in Texas? Sam said no—bad public relations, and ranchers were always in hot water anyway, it seemed. Ranching in general and trapping or hunting in particular can be sore subjects in a modern world that tends to recoil at all bloodshed. Sam's ring of hell would be full of vegan environmentalists. He sees "silly environmentalists" as trying to limit or eliminate his way of life: objecting to predator control, range leases, and other common ranching practices. Many of Sam's battles are physical and direct—encounters with weather, landscape, beasts domestic and wild—and he's not comfortable fighting a distant, abstract adversary. You'd be hard-pressed to find a sweeter soul than Sam's, however little he likes government, taxes, laws, regulations, media—all the outside forces that contribute hardly anything to his daily life except static.

Sam told a good coyote story from his trapping days. He and Kathy were just married, living in the bunkhouse, and drove into Santa Rosa for a movie. On the way back that night, Sam checked a trap he'd set not far from the main road. "Sure enough, I had a coyote," Sam said. "I shot him, tossed him in the back of the pickup, and when we got home, set him at the back door to the bunkhouse. I must have only grazed him—though I shot him in the head and he bled plenty—because when I went back out, he was gone." Among other ploys, coyotes will pretend to be dead either to catch curious game or to escape an enemy. Coyotes play possum. "Well, we decided to put one over on Herman," Sam continued. "I phoned him and told him to please come over, I was in trouble. I sounded real serious. So Herman walks over and the first thing he finds is blood all over the bunkhouse stoop. I mean blood *everywhere*. He walks in and there I sit at the kitchen table, head hanging down, soaked in blood myself. 'Herman,' I said, 'Something terrible happened. Kathy and I were in town, and on the way back home we got into a big fight. I don't know, Herman, one thing led to another, and well . . .' I pointed to the blood all over me. Herman just sat there staring, didn't know what to say. About that time Kathy burst out laughing in the back room. We didn't get the coyote, but we got Herman."

I think about all the elements that come together in that little story: Sam, Kathy, Herman, coyotes, ranch life and death. I can imagine how foreign the experience must seem to most Americans. But hell, after all, isn't this New Mexico?

January 31, 2001—This morning I got up early to make my plane. I liked rising in the dark, as Guy Cowden habitually did, same as his ancestors. The practical reasons for this, doing the work they did in this country, are the same as for siestas. Midday on the plains is no time for strenuous activity by man or animal, so men went to work early, then rested in the heat of the day. Siesta was not a function of Spanish or Mexican or Texan laziness but an appropriate response to the environment. As when a jackrabbit finds shade from a cactus, intelligence is at work. I was reading a prose passage, "Wilderness," in Gary Snyder's *Turtle Island,* in which he discusses the intelligence evident in a grove of trees, an intelligence that reaches back deep into time and incorporates great complexity. Many people can't or won't see that operating, so fixed are they upon themselves and those like them, so fearful are they of truly new experience. What prompted Emerson and Thoreau to face west, away from Europe, toward the dark forest frontier, was dissatisfaction with the comfortable known world. When we face the dark and accompany its change into light, the journey is then complete.

Leaving Albuquerque, we took off north, past the Sandia Mountains, then turned east and passed over the ranch. From my window seat I got as good a look as possible. From thirty thousand feet up, the Cowden Ranch loses its detail; it was difficult to locate except in relation to large and distinctive landmarks. I could see Santa Rosa Lake south of our flight path and, beyond that, the reservoir at Fort Sumner. I saw the Pecos River winding off toward the southeast, joined by the Gallinas, and I located the Conchas River and Conchas Reservoir. Other than bodies of water, only big land-masses were distinctive—river breaks of the Conchas, the big mesa to its north; smaller mesas were difficult to identify. The landscape in general looked to be plains occasionally cut by erosion and drainage into mesas or plateaus. The Caprock of the Llano Estacado to the south appeared as a dark line, a shadow, rather than the well-defined, steep wall that would have confronted a traveler on the ground. The view clearly placed the ranch within a plains, not mountain, environment; the scarcity of water and vegetation was apparent. I see that Guy Cowden did not actually move off the plains where he'd raised cattle all his life; he moved to its margin, another frontier, an edge more accommodating to his way of life. "Same old story. Looking for land," my mother had said of his move away from Midland. "He just wanted a good ranch."

Give Me Land, Lots of Land

[Palo Pinto] was a great cattle range when I first went there. There was lots of water, range, and shelter, and it was all open. . . . Finally smaller outfits began coming in and they crowded the range. The big outfits would never crowd in on another big outfit. When the country began to get crowded, the big men moved on out to the edge of the plains.

Henry Wolcott, in an interview by J. Evetts Haley

These great steppes seem only fitted for the haunts of the mustang, the buffalo, the antelope, and their migratory lord, the prairie Indian. Unless with the progressive influence of time, some favorable mutation should be wrought in nature's operations . . . these high prairies could at present only be made available for grazing purposes. . . . Of this unequalled pasturage the great western prairies afford a sufficiency to graze cattle for the supply of all the United States.

Josiah Gregg, Commerce of the Prairies

The Record of Marks and Brands in the Palo Pinto County Courthouse, Book A, page 16, includes several entries for the Cowdens. In 1872, when they registered their separate earmarks and common brand—did "C. C." stand for "Cowden Cattle" or "Cowden and Cowden"?—the four older Cowden brothers could see their collective future as cattlemen, if not its precise details or dimension. William Henry was nineteen, George Edgar seventeen, John Motherwell fifteen, and Charles Webster twelve. All were experienced cowboys by then, with years in the saddle on cow hunts, helping their father and uncle tend Cowden cattle running in southwestern Palo Pinto County. When the older men made trail drives, their sons likely went along; most cowhands were young. The two baby brothers—Liddon

COWDEN MARKS AND BRAND, PALO PINTO COUNTY, 1872

Number	Name of Owner	Marks	Brands	Location of Brand	Date of Registry
621	W H. Cowdon		C.C.	Hip Shoul	May 18, 1872
624	G. E. "		C.C.	" "	" " "
625	J.M. "		C.C.	" "	" " "
626	Chas. W "		C.C.	" "	" " "
627	C M "		C.C.	" "	" " "

was two years old, Rorie Wynne was one—would be joined in 1875 by another, Eugene Pelham Cowden. Like their father and uncle before them, it was natural for the Cowden boys to work together to make their marks as cattlemen—a tradition of family ranching was firmly in place. W. H. Cowden, Sr., also used the CC brand, on the right and left hip and shoulder of cattle; horses were branded on the right thigh and shoulder; Carrie ("C M Cowdon" in the Record of Marks and Brands) had her own earmarks and share of the Cowden herd. Uncle Frank Cowden, his wife, and children used the FC brand in many different forms, with varying earmarks.

Other early brands included those of Palo Pinto cattlemen Oliver Loving (OL), C. A. Slaughter (CS), J. H. Dillahunty (DIL), C. A. Dalton (DAL), W. C. McAdams (WM), J. A. Lynch (JL7), among many others. Brands were registered county by county; in *Texas Cattle Brands*, Gus Ford estimated that some 500,000 brands were registered in the first hundred years of Texas history. Brands were sometimes bought and sold with the cattle wearing them, or cattle might wear two or three brands if they had changed hands. And since many different owners might contribute to a trail herd, drovers almost always added a trail brand. On their home ranges, Palo Pinto residents typically followed the custom of their southern Anglo-American ancestors, using block capital letters from their names and, sometimes, numbers. In contrast to this tradition, brands used by Spanish and Mexican ranchers tended to be abstract and often elaborate: "crazy Mexican brands . . . with complicated crooked and curved lines and many curlicues." Ford suggested it was "the push to the West and the broad expanse of prairie that made bold brands, lavishly displayed, necessary," noting that "the cowman of the 70's and 80's wanted and needed a brand that he could see at a distance and one which discouraged reburning by the 'light-fingered gentry.'" The JAL brand made famous in New Mexico and West Texas by the Cowdens would eventually produce confusion about its origins in Palo Pinto.

As the Cattle Kingdom expanded from Texas north and west, dispersing herds and cowboys and ranching practices across the Great Plains from the Rio Grande all the way to Canada, Palo Pinto and other northwestern counties saw the frontier extend beyond their borders. Wagons loaded with great stacks of shaggy buffalo hides began to arrive from deadly-efficient hunters out on the plains—an estimated 4.5 million buffalo were slaughtered from 1873 to 1875 for their hides alone. Grazing lands opened before cattlemen at the same time they felt increasing pressure from new settlers moving in. The 1860 census figures for Texas show Palo Pinto County with a population of 1,524; because of Indian troubles, no tally

PALO PINTO COUNTY BRANDS

Oliver Loving

James Alonzo Edwards,
to Cowden brothers

W. W. Cochran

Trail Brand,
Loving and Goodnight

C. A. Slaughter

SPANISH BRANDS IN TEXAS

Don Juan Joseph Flores

Jose Maria Cortinas

Jose Antonio Navarro

Manuel Sanchez

Domingo Gonzales

was made in 1870, but the population on the frontier tended to decline for the decade. By 1880 the population had increased to 5,885; the census for that year listed 648 farms and ranches, with 42,400 cattle and some 5,000 sheep. Farmers raised corn, cotton, and wheat, mostly in the eastern part of the county. There the town of Mineral Wells arose at the site of a water well drilled by James A. Lynch, soon touted for its medicinal properties—it contained lithium, Epsom salts, and other treats. To those accustomed to space and solitude, the world must have seemed diminished and crowded, no matter that new settlement brought economic opportunity. In 1876 the county seat was "a thriving frontier settlement and big ranching center," according to the *Palo Pinto County Star*. "It was common to see huge herds of longhorn cattle driven through the town daily on the trail to market. Wagon trains crept along the dusty, rocky roads to the forts on the frontier border and the old stagecoach drew up on schedule in clouds of dust." Growth was clear from the merchants joining Dillahunty; J. B. Fleming, G. W. McDonald, J. W. Schoolcraft, and H. G. Taylor all ran stores. A grocery and numerous saloons sprung up, as noted in the brand-new *Star*: "Just received at the Star Saloon, a lot of fine Tennessee and Kentucky whiskies, and the genial proprietor, Captain Hamilton invites all to call and try them for medical purposes or otherwise, as they are pure and unadulterated." Schoolcraft's store advertised the following prices:

Corn per bushel	.90
Oats per bushel	.75
Flour per hundred	4.75
Apples, dried, per lb.	.12
Peaches, dried, per lb.	.12
Butter, per lb.	.20
Eggs, per dozen	.12½
Lard, per pound	.15
Sugar, per pound	.12 to .15
Coffee, per pound	.24 to .25
Tea, per pound	1.00
Rice, per pound	.10
Bacon, per pound	.12½
Chickens, per dozen	2.00
Prints, per yard	.07 to .08

Out on their ranches, the Cowdens were relatively self-sufficient yet aware of developments in town and throughout the area: merchants and farmers moving in, new services and goods available. One item in particular was of interest: wire fencing material, made of steel with sharp barbs woven or welded to the strands of wire. A farmer named Joseph Glidden up in DeKalb, Illinois, had filed a patent on his invention in 1874 and started selling it the next year, at two cents a pound; by 1876, barbed wire was being pitched in Texas. Glidden's wire was featured in a demonstration for skeptical ranchers and farmers down in San Antonio, where half-wild Longhorns were released into pens built of "Winner" wire, which effectively and marvelously held them in place. "Light as air, stronger than whiskey, and cheap as dirt," barbed wire was a revelation. Now a man could fence his land, even where there was little timber. Rock or wooden picket pens, common in Palo Pinto and the Cross Timbers, were fine for milk cows or for working cattle—they had been used for decades throughout the South— but impractical for fencing fields or rangeland. In Texas, cattle were free to range anywhere, a practice that arose from Spanish and Mexican open-range ranching rather than from British traditions in place back East. Since cattlemen weren't required to control their stock, a Texas farmer who wanted to protect his crops had to fence cattle out. With barbed wire and a few cedar posts, a man could fence his whole farm. And what with homesteaders moving in, cattlemen themselves might use "bobwire" fences to establish their ranges, although many disliked any constraints on open ranges. Glidden and other manufacturers of the new wire would prosper as homesteaders and ranchers collided on the old frontier and as cattlemen moved operations out onto the treeless Great Plains, where steel wire could bring hundreds of thousands of acres of grazing land under control. Glidden's partner, Isaac Ellwood, the original inventor of "thorn wire" fencing, eventually bought the famed Spade Ranch in the Texas Panhandle and eventually lamented its breakup into parcels of farmland— a casualty of his own invention, among other pressures upon the open range.

As the Brazos country began to fill up with new settlers, the Cowdens also came face-to-face with their own increase. Bill and Carrie had brought twelve children into the world, and now their futures were at stake. The Cowden boys no doubt looked around and wondered just exactly where they might run the herd of cattle they were building. Well-watered grazing land along Rock Creek and other Palo Pinto tributaries was getting scarce. In the same position were Frank Cowden and his wife, Elizabeth, who had raised nine children—six sons, three daughters. Cowdens abounded in Palo Pinto; the combined forces that provided security and companionship on the early frontier now threatened to overwhelm them.

In the midst of this progress and plenty—Indian menace apparently over, trails and markets for cattle now available, big herds and big families—came what had to be a transitional moment for the family. On September 16, 1879, Carrie Liddon Cowden died. Forty-seven years old, mother of twelve children ranging in age from twenty-eight to four, Carrie had lived on the harsh Texas frontier most of her life—such strains took an early toll. She was buried near the South Fork of Palo Pinto Creek in Davidson Cemetery, next to her infant daughter and her firstborn child; a third daughter would die the next year. The deaths of four Cowden females lent force to the old saying that "Texas is fine country for men and dogs, but hell on horses and women." A family history contains a tribute to the pioneer mother and wife of W. H. Cowden: "Of his devoted wife, his counselor and sharer of joys and sorrows, a volume might be written. She met all things as they came, with calmness and courage. She endured hardships without a murmur, and gave her life to the raising of her family, to the training of her children in 'the way they should go' and in the belief and practice of the Christian Faith, giving them of her own fine nature. At length the Call of the Master came, and she laid down the life so full of hopes, and duties, and passed on to the Higher Life."

Hopes and duties remained for the family, particularly the Cowden boys. In 1875 the oldest son, Bill Cowden Jr., established his own ranch nearby on Walnut Creek. After his mother's death in 1879, Bill Jr. and his brothers moved seven hundred head of Cowden cattle west a hundred miles to Paint Creek, territory Comanches had ruled not long before. They returned to Palo Pinto the next year, but clearly they were looking for new ranges. Carrie's death may have helped to sever ties to the country where the Cowdens had lived for twenty-five years; the marriage of Bill Jr. to Mary Salvage in 1880 provided additional motivation to the restless ranching family. The next generation was ascending.

Also in 1880, the Texas and Pacific Railway Company, operating with a federal charter and a mandate to reach California, began extending its line from Fort Worth west toward El Paso. Not built to serve existing populations—only a handful of frontier forts and towns could be found for six hundred miles—but designed to create a corridor of development across the empty Texas plains, the railroad was promised twenty sections of state land for each mile of track constructed. This policy was a mixed blessing for cattlemen in its path. They would have transportation to markets, and supplies could be brought in readily, but their ranges would be compromised, their homeland and herds cut up by encroaching development. The arrival of the Texas and Pacific Railway in Palo Pinto gave the new generation of Cowdens one final motivation to depart.

When I consider why the Cowdens picked up and moved from a well-established life in Palo Pinto—another westward leap—a couple of stories help reveal their state of mind at the time.

The *Dallas Morning News* once ran an article about the "Million Dollar Cow" of Palo Pinto. In 1880 when the Texas and Pacific was laying track, the line headed due west from Fort Worth to Weatherford, east of the Palo Pinto County line. Like frontier folk elsewhere in the path of railroads, tradesmen in the county seat of Palo Pinto eagerly anticipated the bonanza the railroad would bring once it came to town. Their fortunes would be made, they thought.

But as surveyors and track engineers were working their way west, figuring the route, plans changed. Railroad crews were in the habit of supplying themselves with beef from the ranges they passed through. They'd kill a cow and send the hide with the brand into the T&P office, which would identify the owner and draw up a check to pay for supper. It was a common practice, but one unfamiliar to Palo Pinto cattlemen. Frontier ranchers were skeptical to begin with, unaccustomed and resistant to intrusions upon their ranges, even from a railroad that might guarantee them access to markets and financial success. They were independent sorts, used to trailing cattle across open country to markets—the freedom of that practice was more appealing than the convenience of a railroad at their front doors.

So when the railroad men killed the Palo Pinto Longhorn—worth at the time about ten dollars in Texas and perhaps forty dollars up in Dodge City, Kansas—the owner took great exception. With his rifle he rode to the railroad camp, where he found T&P men, armed themselves and in no mood to entertain blistering complaints. The cowman returned to Palo Pinto, assembled an armed party, and a heated standoff followed between cattlemen and railroad crew. The surveyors reported back to the office that Palo Pinto ranchers didn't want railroad crews butchering their cattle, didn't want any damn railroad running through their ranges. To this ingratitude the Texas and Pacific brass responded that engineers should "build around the troublemakers—even if it takes a million dollars." Thus the railroad swung south at Weatherford, bypassing most of Palo Pinto, leaving the county seat high and dry as the tracks angled southwest, clipping only the lightly populated country along Palo Pinto Creek.

Then, into this climate of confrontation rode Jefferson Davis Cowden. Jeff Cowden was first cousin to my great-grandfather, one of the host of young Cowden cowboys riding the ranges in 1880. Although he would eventually become a Texas Ranger and a U.S. marshal, like most frontier cowboys and most Texas Rangers, Jeff was a rough customer. He was riding with his cousin T. E. Bell along the railroad right-of-way, where crews of Irish laborers were laying track, when a train approached from the east with flatcars of supplies—rails, ties, spikes, and goods for the crews. Jeff looked at the steam engine smoking through his cattle range and suggested to his companion, "Ed, nothing like that can go through this county without being roped."

"Go to it," Bell replied. "If you don't, I will."

In the brazen way of cowboys, Jeff Cowden readied his rope, spurred his cow pony up alongside the Texas and Pacific engine, and threw a perfect loop over the smokestack. Leaning on picks and shovels and sledgehammers, the railroad crew watched the drama unfold, cheering the cowboy on. What they didn't know was that for most cowboys it was a point of honor that they never lost a rope, never cut themselves and their horses free from whatever they caught—outlaw Longhorn, mustang, grizzly bear, steam engine. Texas cowboys at that time tied their ropes "hard and fast." So Jeff Cowden found himself galloping alongside an impossible prize, unable to bring it to heel and unwilling to back off, until the T&P engineer gave an extra blast of steam to his engine and burned Jeff's loop free from the smokestack.

Such is the stuff of family legends. My mother tells the tale, which she heard in high school as part of a Texas history lesson, with embarrassment. Her family, those wild cowboy Cowdens, roping railroad trains! She remembers: "I buried my head in my arms on my desk—mortified."

The Irish laborers, however, took great delight in the cowboy who lassoed the locomotive, and the story spread, eventually reaching the offices of the Texas and Pacific. After their recent confrontation with angry Palo Pinto cattlemen, the T&P executives apparently chose to view Jeff Cowden's exploits as a public relations opportunity rather than as more bad behavior by frontier cowboys. They decided to acknowledge his dubious achievement with a lifetime pass on the Texas and Pacific Railway, helping defuse the tension of its intrusion upon the open range.

To the annals of the "Wild West" was added a Cowden tale, an adventure that shows not only the independent spirit of frontier cowboys but also their tenuous position in history. Jeff Cowden roping the T&P steam engine now seems much like Quanah Parker's Comanches attacking the deadly buffalo guns at Adobe Walls in spring 1874. Inspired by a medicine man who assured them that they were invulnerable and could sweep away the whites slaughtering their buffalo herds and settling their river valleys and grasslands, Comanche, Kiowa, and Cheyenne warriors made one final attempt to assert control over the plains. They massed around an isolated buffalo hunters' camp on the Canadian River in Texas for a dawn attack, which the hunters accidentally discovered, barricading themselves behind the camp walls and using their long-range Sharps rifles—known as the "gun that shoots today and kills tomorrow"—to devastating effect. Twenty-eight hunters fought off seven hundred warriors. Their attack repulsed, the Indians were forced to retreat before superior firepower and unlimited reserves of manpower inevitably heading west. When the U.S. Army that fall caught up with the "renegade" Comanches and slaughtered their entire herd of fifteen hundred horses—as central to their life as buffalo—all was lost.

Comanches and buffalo guns, cowboys and locomotives—the trend was clear in acts of heroic but doomed resistance. Technology would triumph over spirit; "progress" would prevail. Jeff Cowden's moment of glory was history.

After the railroad arrived, the little towns of Santo, Gordon, Mingus, and Strawn began to take shape beside the tracks, which paralleled Palo Pinto Creek. The next October, 1881, the Cowdens took their herd west once again, more than 150 miles to the Double Mountain Fork of the Brazos, north of Colorado City, Texas. Their new range had been described by Captain Randolph Marcy on his return in 1849 from Santa Fe, scouting a route for an eventual railroad line across the southern plains: "We came upon the high banks of the Double Mountain fork of the Brazos, or 'Tock-an-ho-no,' as the Comanches call it . . . [and] continued to the south, crossing a spring branch, and passing over as beautiful a country for eight miles as I ever beheld. It was a perfectly level grassy glade, and covered with a growth of large mezquite trees at uniform distances, standing with great regularity, and presenting more the appearance of an immense peach orchard than a wilderness. The grass was of the short buffalo variety, and as uniform and even as new-mown meadow; and the soil equally as rich, and very similar to that in the Red river bottoms. This, together with the fact of its being well watered with small spring brooks, gives it all the requisites for making beautiful plantations that the most fastidious amateur in agriculture could desire."

Such promising territory had drawbacks, as Marcy's journal details over the next few days, in mid-October: "We have had during last night one of the most terrific storms I have ever witnessed in the whole course of my life. The wind blew a perfect tempest from the north, and it appeared as if the whole flood-gates of the heavens were suddenly opened, and the accumulated rains of a year poured out in torrents for fifteen consecutive hours upon us. . . . The creek upon which we are encamped had but very little water in it last night: it is now full to the top of its banks, and would float a steamboat." A few days later, yet another norther blew down upon Marcy's camp: "Last night was one of the coldest I have ever known at this season of the year. About dark the wind turned to the north, bringing clouds and rain, and this morning the surface of the ground is covered with snow."

Mercurial weather on the southern Great Plains did not deter the Cowdens, who nonetheless would not remain on the Double Mountain Fork but would move their new cattle range farther west. Likely they wanted to be well beyond the reach of other cattlemen and new settlers, in country they could have to themselves.

In 1882—the year Bill Cowden, Sr., remarried, to Carrie's widowed sister, Kitty—the Cowden boys made a historic purchase. They bought one hundred head of cattle from Bill Jr.'s father-in-law, James Alonzo Edwards, who had registered the JAL brand in Palo Pinto County on May 6, 1872. "Lon" Edwards and his brother Jerome ran cattle in the southwest part of the county, about ten miles from the Cowdens, whom they had joined in the past to put together trail herds, once taking cattle as far as Oregon. On their return, Cochran reported, "Jerome and Alonzo sold their ED and JE brands of cattle," at which point "Alonzo started the famous JAL brand." Why "JAL"? Cochran did not address the choice; perhaps it represented *James Alonzo* or *Jerome And Lon*. To avoid blotching that occurred when branding irons were formed with crossing lines, the A in JAL was an "open" letter, with no crossbar.

The origin of the JAL brand is clear, despite later speculation and claims, as in *The Handbook of Texas:* "The brand's origin remains a mystery; theories claim that it came from the initials of John A. Lynch, James A. Lawrence or John Allen Lee, all of whom were early West Texas ranchers." Palo Pinto County records do not show, however, whether the Cowdens bought both cattle and rights to the brand, which was common. In any case, the hundred head of JAL cattle were thrown in with the rest of the Cowden herd out on Double Mountain Fork, five counties west, where rights to the Palo Pinto County brand did not apply. This gave the family more than eight hundred head, the foundation herd for the JAL Ranch in West Texas and New Mexico, where they would soon move.

Cattlemen from the early Palo Pinto days were scattering across the new frontier. Charles Goodnight was long gone, first to Colorado, then back to the Texas Panhandle in 1876. The next year C. C. Slaughter moved a herd out near Big Spring, along the upper Colorado River. Like Goodnight and Slaughter, other enterprising, independent cattlemen were looking for ranges on the rolling prairies of Texas and beyond. Standing always in the way of direct westward movement was the Llano Estacado, the Staked Plains, which began at the rugged Caprock escarpment, rising three hundred to a thousand feet from the base to the level grasslands above.

American settlers now confronted from the east the same impediment Spaniards had faced from the west for three hundred years. From its beginning in the north near the Canadian River, the western Caprock—sometimes called La Ceja, Spanish for "the Eyebrow," for its fringe of evergreens atop the escarpment—paralleled the Pecos River down through New Mexico. The *ceja* was the *estacado,* the "palisade" that Coronado's expedition had surmounted centuries before. But many explanations are given for the curious name. According to Herbert Bolton, "Llano Estacado" was "later mistranslated by Anglo-Americans into 'Staked Plains,' which completely missed the point of the Spanish designation. They were called Stockaded Plains from the rim-rock, which at a distance looks like a stone fortification." *Estacado* is the past participle of the Spanish *estacar,* "to put stakes in the ground, to enclose with stakes." In myth and perhaps in fact, the Staked Plains were so named because, on the featureless surface, stakes were driven to find the way back from journeys—a task otherwise difficult or impossible. Other explanations of the "stakes" include tall stalks of yucca that arose here and there from the grassland, or the pickets used to tether horses, since no trees were available, as one Spanish explorer reported: "There was nothing but grass and a few small pools of rainwater, with very little water, and some dry holes; on the plains where we camped it was necessary to drive stakes for the guard and reserve horses." Indians guiding Coronado shot arrows before them, following the trail from shaft to shaft across the plains—another popular explanation. Historian H. Bailey Carroll suggests that "the original term might have been *Llano Destacado.* The verb *destacar* from which the participle *destacado* is formed means to detach from the main body of which it is a part, but the word is commonly used to mean to emboss, to elevate, to stand out, or to raise." In lazy frontier Spanish of that time, the *D* was dropped, leaving the Llano Estacado mesa to loom before travelers then and historians now.

Whatever its metaphorical source, the Llano always intrigued explorers: "We reached some plains as bare of landmarks as if we were surrounded by the sea. Here the guides lost their bearings because there is nowhere a stone, hill, tree, bush, or anything of the sort. There are many excellent pastures with fine grass." Coronado's letter to the king of Spain was the first of many accounts, followed centuries later by that of American explorer Josiah Gregg, who crossed the Staked Plains to and from Santa Fe in 1839 and 1840: "The most notable of the great plateaux of the Prairies is that known to Mexicans as *El Llano Estacado,* . . . quite an elevated and generally a level plains without important hills or ridges, unless we distinguish as such the craggy breaks of the streams which border and pierce it. It embraces an area of about 30,000 square miles, most of which is without water during three-fourths of the year."

Topographical engineers of the U.S. Army later surveyed the area, reporting, "The formation of the Llano Estacado is one of the most marked physical features of the American continent. Its surface, rising over a broad area to an altitude, in almost every part, of over 4,000 feet, at the lowest estimate, and but little broken or traversed by river valleys, constitutes one of the most perfect examples of an elevated plateau, or *mesa,* that is found. . . . [T]here is nothing to break the monotonous desert character of its surface, except an occasional river gorge or cañon, invisible from a distance, and often apparent only when the traveller stands on its brink."

The southernmost extension of the High Plains, El Llano measures 300 miles from the Canadian River down beyond Midland in the south and 150 miles from the Pecos River to the eastern Caprock. The flat surface slopes imperceptibly southeast—treeless, windblown, and desolate when first encountered by Americans. A romantic early description by traveler Albert Pike conveyed its emotional effects: "Its sublimity arises from its unbounded extent, its barren monotony and desolation, its still, unmoved, calm, stern, almost self-confident grandeur, its strange power of deception, its want of echo, and in fine, its power of throwing a man back upon himself and giving him a feeling of lone helplessness, strangely mingled at the same time with a feeling of liberty and freedom from restraint."

Rather than ascend onto the open, unprotected, "waterless" Llano, Anglo cattlemen began to establish themselves at its edge, a fringe of rugged canyons formed by headwaters of rivers running east and south across Texas: Palo Duro Canyon on the Prairie Dog Town Fork of the Red River, Quitaque Valley on the Pease River, Blanco Canyon on the White River, and Yellow House Canyon on the Double Mountain Fork of the Brazos. Buffalo herds and Plains Indians had frequented these canyons for generations, and they were ideal cattle ranges—well watered, providing protection from winter storms sweeping down across the Great Plains, and helping ranchers to control movement of their herds. Charles Goodnight was one of the first to recognize the potential of this natural arrangement; with help from the Irish investor John Adair, he established the JA Ranch in Palo Duro Canyon. Other Texas cattlemen and foreign investors soon moved to the hospitable borders of the Llano Estacado. George Littlefield's LIT began running cattle in 1877 up on the Canadian River to Goodnight's north; Littlefield would later acquire additional ranchland across the Llano, along the Pecos River in New Mexico. In 1878 the Matador Ranch was established to Goodnight's south, on the Pease River headwaters; that same year the Spur Ranch set up operations in Blanco Canyon to the south. By 1882 the Yellow House Ranch was operating on the headwaters of the Double Mountain Fork, so that the principal rivers—what was called, along with the isolated permanent springs, "living water"—were all occupied.

The Cowdens' old Palo Pinto neighbor, C. C. Slaughter, was established on the Colorado River to their immediate west, so to find wide-open country, uncrowded and unlikely to become crowded with people or cattle, they looked beyond the eastern borders of the Llano, toward the sandhills west of Midland. As Walter Cochran remembered, "The JAL Ranch was started in the fall of 1883. W. H., George, John M., and Buck [Charles Webster] Cowden had gone from Palo Pinto County and wintered on the Double Mountain Fork of the Brazos. From there they moved to the White Sand in December, 1883. They had about 1000 stock cattle with them and theirs was the first ranch in that country. There was living water in the Sand when they went there and this was the only water in the country."

Ranching in the sandhills of West Texas and New Mexico had never been attempted. For decades this striking geological feature had intrigued and intimidated travelers; after Marcy traversed the northern Llano Estacado in the spring of 1849, he returned along its southern perimeter in September, recording one of the first American views of the area: "The whole surface of the country . . . seemed to be one continuous succession of white sand hills, from twenty to one hundred feet high, in which [our] horses sunk to their knees at almost every step. . . . These hills, or mounds, present a most singular and anomalous feature in the geology of the prairies. They extend (so far as we have explored) at least fifty miles in nearly a north and south direction, and from five to ten miles east and west; they are white drift-sand thrown up with much uniformity into a multitude of conical hills, destitute of soil, trees, or herbage." Later that year Lieutenant Nathaniel Michler followed Marcy's route from the east: "Then come the white sand hills, which are really an object of curiosity. They are a perfect miniature Alps of sand—the latter perfectly white and clean; in the midst of them you see summit after summit spreading out in every direction, not a sign of vegetation upon them, nothing but sand piled upon sand. They form a belt two or three miles in width, and extend many miles in a northwest direction. But a matter of the greatest surprise is to find large water-holes among them; they are found in the base of the hills, are large, deep, and contain most excellent water, cool, clear and pleasant. The water is permanent."

Marcy's southern path became known as the Emigrant Trail, leading travelers to the gold fields of California; the lower Staked Plains were an

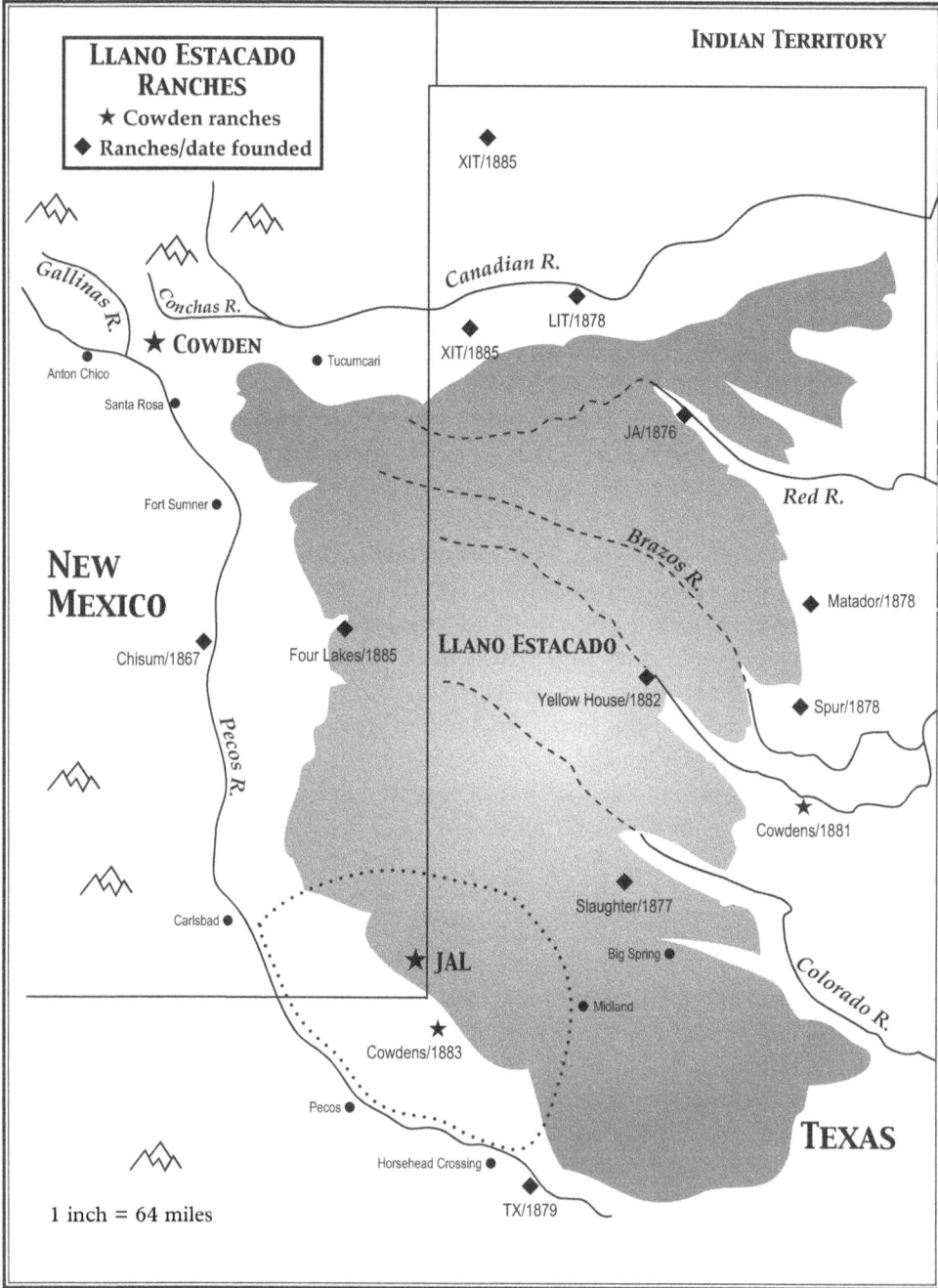

LLANO ESTACADO
RANCHES
★ Cowden ranches
◆ Ranches/date founded

INDIAN TERRITORY

Gallinas R.

Conchas R.

Canadian R.

★ COWDEN

● Anton Chico

● Tucumcari

XIT/1885

XIT/1885

LIT/1878

● Santa Rosa

JA/1876

Red R.

● Fort Sumner

NEW
MEXICO

Brazos R.

● Chisum/1867

◆ Four Lakes/1885

LLANO ESTACADO

◆ Matador/1878

Yellow House/1882

◆ Spur/1878

Pecos R.

★ Cowdens/1881

● Carlsbad

Slaughter/1877

★ JAL

Big Spring ●

Colorado R.

● Midland

★ Cowdens/1883

● Pecos

TEXAS

Horsehead Crossing ●

TX/1879

1 inch = 64 miles

obstacle on the way to promised lands—no one took them for a final destination, a home of any sort. Five years after Marcy's and Michler's journeys, private and government survey parties crossed the sandhills near the 32nd parallel, looking for a route suitable for a railroad to the Pacific. The U.S. secretary of war at the time, Jefferson Davis, favored a southern route over others being considered along the 35th parallel (along the Canadian River), the 40th parallel in Kansas (the eventual transcontinental railroad line), and the 45th parallel. Davis appointed Captain John Pope of the Corps of Topographical Engineers to survey the 32nd parallel in the spring of 1854; in January, surveyor A. B. Gray had led a similar expedition funded by private interests. Gray's account reinforces how strange the country appeared: "These singular-looking hills seemed to be an accumulation of fine white sand, heaved together near the lower part of the Staked Plain; extending south-easterly for fifty or sixty miles from the parallel of 32°, with an extreme breadth of ten miles; and innumerable hillocks and ridges 40 to 80 feet high, that at night resemble ocean waves. Though fatigued by a long march, there was something exceedingly interesting in our passage through this granular sea, the alternating light and shade, occasioned by clouds passing before the moon, gleaming of the water, and the uniform ripples in the sand, adding to the general beauty, and exciting our wonder and admiration."

In May 1999, I visited Monahans Sandhills State Park, near where the Cowdens arrived with their JAL herd in 1883. What possessed them? I know that water—so rare and valuable in the Southwest—drew them to the sandhills, which makes the "desert" landscape even more surreal. I exited I-20 and drove into another world, dune upon dune of white sand through which the park road wandered, placing and displacing me at once. The landscape overwhelms memory and imagination, insisting upon its own particulars—dazzling sunlight upon sand, occasional clusters of low shin oak, wind blowing sand from a dune ridge, a single cloud drifting across sharp blue skies—yet I half expected sounds of the ocean rolling in nearby. Nothing. I was fifteen hundred miles from the Atlantic, a thousand miles from the Pacific, five hundred miles from the Gulf of Mexico. I was at the junction of the Great Plains and the Chihuahuan Desert—I heard nothing but dry wind.

I took in a geology lesson, walking through the park museum with its charts and photographs and diorama, information conveniently collected for me more than a century after the Cowden boys decided that these sandhills were the perfect place for them and their JAL herd. Well, perhaps not perfect. But they were free—no one else had claimed the land. For all of recorded history, no one had settled here permanently.

I learned that the sands are new, less than twenty thousand years old. Many millions of years before, the Rockies had formed to their northwest; from those steep mountains, rivers like the Pecos carried runoff east and south across the southern High Plains. When the last ice age ended a mere million or so years back, the climate in the Southwest grew drier and warmer, until many rivers failed to run permanently. Then the wind took over, paleowinds blowing relentlessly from the northwest, carrying sand and silt from dry streambeds, depositing them along the southwestern edge of the emergent Llano Estacado. Dunes formed, drifting back and forth as winds eventually shifted with the seasons: winter storm winds from the northwest, transitional spring winds from the southwest, steady southeast winds during summer. It's as if a great hand slowly stirs the landscape. Over time, during periods of greater rainfall—called pluvial periods—tough vegetation took hold among the sandhills and helped stabilize them, reducing the amount and degree of movement. The sandhills now have active (moving) and vegetated (stable) dunes, both subject to seasonal winds and change. Time, water, and wind are the great forces behind the seas of sand.

Here the Cowdens made their first stand in far West Texas, setting up tents like bedouins on the Sahara, then building dugout homes for their families on "hard land" nearby. The sand was a fact of life, sand and wind and sky their world, just as other pioneers had woods and streams and meadows. My mother tells me that when she was a child on the ranch, fifty years after the Cowdens arrived, they still lived in and with the sand. For fun they'd go on picnics in the sandhills, sliding down them like other children sledding over slopes of snow, or sitting on blankets as other families might on ocean beaches. Toward every horizon stretched the white dunes.

And my great-aunt Olive Cowden told me a story of a wild ride with her husband and infant daughter on a mule-drawn wagon through the sandhills. "Al and Emma Sue and I were going across the sand, going to check one of the windmills. It was 1932. We had Old Red and that other mule . . . I forget what the name of the other one was, but he was real contrary. Old Red was full of life and vinegar. Well, we were in the deep sand, and all of a sudden the mules went to snorting with each other. And directly they went to running, so Al said, 'Mama, jump out quick! They're going to run away.' He always called me Mama.

"So I said, 'Al, will it hurt Emma Sue?' I couldn't imagine jumping.

"He said, 'No! Jump!'

"So I held her and jumped. Jumped off the racing wagon with my baby girl. I look back over that and think, 'My goodness, how did I do such a thing?' Because I might have killed her! But that deep sand was as soft as could be—neither of us got a scratch. And Al finally got the mules under control without them turning over the wagon. That was one time I gave thanks for the sand."

Bessye Cowden Ward also remembered ranch life in the sandhills: "We were at the ranch north of Kermit, and me just a little girl. We'd go in the Model T Ford, and the sand would be so bad in the ruts of the road—which was just a wagon track road for about a mile or mile and a half nearly—and Daddy would saddle a horse. . . . And he'd tie a rope to the horn of the saddle and to the radiator cap of the Model T Ford. And he would drive the Ford through, 'course just in low gear, all he could give it, and I would pull with the horn of the saddle. I'd ride the horse and pull to get this car through the sand. We would. I'd get off then and unsaddle, put the saddle under a mesquite bush, and tie the horse to the mesquite bush with a long stake rope and go on in and get our groceries and come back and saddle the horse and tie on the car just the same way to take it back to the house. And that was done frequently."

And this was the land, this was the life, the Cowdens chose.

The Pacific Railroad surveys, and an attempt by Captain Pope to drill artesian wells near the Pecos during 1855–57, did not result in immediate Anglo-American movement into West Texas. Emigrants to California hurried through and cattle drives swung along the Pecos up into New Mexico, but the formidable landscape and threat of Indian hostilities kept settlers away. And once the Civil War began, no federal resources were available to develop railroad lines or other stimulants to settlement. The first significant government activity in the southern Staked Plains occurred after the war, in 1871, when Lieutenant Colonel William Shafter led troops of the Twenty-fourth Infantry and Ninth Cavalry—Negro regiments, storied "buffalo soldiers"—in pursuit of Comanches who had stolen mules and horses from the army. Shafter reported on the sandhills and Pecos plain: "I have never seen better grass in Texas than was to be found in all the valley we passed through and the water in the White Sand Hills is permanent. Willow trees ten inches in diameter are at the edge of the water holes. These hills are of fine white sand and in every light wind it ripples like water over the sides and tops of the hills. There is no grass in the hills but it is but a mile or two either east or west to the prairie where the grass is excellent." Shafter and his troops spent three weeks tracking

Indians, circling through the sandhills and lower Llano Estacado. After following Monument Draw up into New Mexico, passing the site where the Cowdens would establish headquarters for the JAL Ranch, Shafter's troops came upon an abandoned Indian camp. Like the water and willows Shafter had seen, the presence of Indian camps—both Comanche and Apache—spoke to the value of the sandhills as, ironically, an oasis in the semiarid country.

In 1875, Shafter made another excursion through the sandhills and onto the Llano Estacado. For almost six months he led Negro infantry and cavalry, Seminole scouts, teamsters, surgeons, and civilians—a force of five hundred—on a twenty-five-hundred-mile expedition over the lower Staked Plains. His report proved valuable to cattlemen looking for new ranges; his route crossed the Double Mountain Fork of the Brazos, where the Cowdens would later graze their herd. "The whole country is covered with luxuriant grass, affording pasturage for immense herds of buffalo," he wrote, "and would be sufficient to maintain thousands of cattle and horses." Shafter also returned to the sandhills: "A range of low hills of very white sand, without vegetation, and almost impassable, except for horses; at least double teams would be required to draw lightly loaded wagons through them. They present, from the distance of a few miles, the appearance of hills covered with snow. . . . Water in almost unlimited quantity can be had by digging in the small depressions at the bases of the hills at a depth of two to four feet. . . . There are also quite large willows and cottonwood trees growing in them, a sure indication of living water. . . . The country east of the sand hills to Mustang and Sulphur Springs is high rolling prairie, covered with fine grass, has no known living water but [an] abundance during the rainy season in small lakes."

The "rainy season" in West Texas and southeastern New Mexico is often little more than spotty, if sometimes severe, thunderstorms in summer; rainfall averages less than thirteen inches per year. But rainwater—stored between the loose grains of dune sand held up by deeper, impenetrable deposits of caliche (lime-rich hardpan) and clay soils—is readily available in the sandhills. After heavy rains, surface ponds are visible, and except in times of extreme drought, water is not far beneath the sand; naturalists have observed coyotes using their claws and teeth to dig for water, just thirty inches down. Evaporation due to heat and wind would measure over thirty-three inches per year without the protective cover of the sand—the forbidding sandhills are actually an efficient reservoir of water. It's like ice conserving heat in an igloo.

Growing down to the subsurface moisture are the willows that Shafter observed around water holes; mesquite, yucca, and cactus also dot the dunes. But the dominant plant in the sandhills is shin oak (*Quercus havardii*), also called shinnery. Although shin oaks grow only about three feet tall, they can have roots longer than seventy feet and often form a dense dwarf forest. The shinnery helps to stabilize dunes, but more importantly for the Cowdens and subsequent cattlemen, the oaks produce a host of large acorns on which cattle browse and thrive. Before and after Anglos arrived with their herds, shinnery helped sustain deer, coyotes, foxes, bobcats, possums, javelinas, porcupines, skunks, rabbits, and, of course, squirrels. Wild game and water, as well as acorns and mesquite beans for meal, had made the sandhills a productive camping site for Comanches and Apaches for hundreds of years. "There were plenty of Indian signs in the way of old camps and trails leading to the Pecos, and there were still a scattering of buffalo in the Sand Hills and antelope by the thousands," wrote W. C. Cochran. Cowboys followed Indians into this odd oasis surrounded by endless grasslands.

Aware of the reports of reliable water and open rangeland, the four eldest Cowden brothers—Bill Jr., George, John M., and Charles—headed for the sandhills. "W. H. Cowden had heard old Matt Tucker of Stephenville [back in the Cross Timbers] tell about tying his mule to an old willow tree in the Sand Hills on his way to California in 1849," reported Cochran. "He hunted and found the tree where Tucker had tied his mule, and made his first camp." The Cowden men trailed cattle, their wives drove wagons of supplies and a few possessions, and the families arrived "with nothing but the shirts on their backs, a few cows and many children." The move echoed that of their father and uncle to the Palo Pinto frontier thirty years before, this time with more Cowdens and more cattle and to a more demanding landscape.

They began to settle in, although conditions were hardly settled on the open range. Most of the Cowdens lived those first years in wagon beds or tents, but George built a dugout house on the edge of the sandhills. It seemed like reasonable, even luxurious, shelter for his family . . . until their milk cow, grazing on the sod roof, plunged through down into the bedroom. George dug away a wall to get the cow out and his family back inside. Rather than building houses—no wood was available anyway—the Cowden brothers built up the JAL herd, tending their wide-ranging cattle, spending their lives in the saddle, almost as nomadic as Comanches who had followed buffalo over the plains for generations. The Cowdens, along

with in-laws W. C. Cochran and Spence Jowell, finally built working headquarters for the JALs on Monument Draw in 1886—a three-room structure
called Old Adobe. Most "business" was conducted in Midland, a seventy-
five mile ride, much of it through the sand, and trips to town were rare, for,
as Cochran reported, "it used to take three to five days in an old wagon to
make a trip to town after a little grub for any of the ranches."

One early task was to dig wells, to provide additional water and extend
their range. "In 1885 they moved up JAL Draw and went to digging wells
by hand, over which they erected windmills," recalled Cochran, who that
year moved from Palo Pinto to join his old neighbors. The Muleshoe brand
from Palo Pinto accompanied Cochran, and his "Muleshoe Watering"
would eventually become the town of Jal, New Mexico. Other waterings
like the Upper Jal, Bullhead, West Well, Deep Wells, and the 84 Well were
created roughly ten miles apart, mostly along Monument Draw, a dry
watercourse that runs east and south from Monument Springs. "Those
were the first windmills in that country," Cochran noted. "They moved up
there to get grass. Their cattle ran all the way from Midland to Carlsbad
in New Mexico, and from there to Horsehead Crossing on the South." If
not overstated, the area that Cochran described—the JAL range at its
widest extent—is approximately the size of Connecticut and Rhode Island
combined, almost 350 miles in circumference, "enclosing" more than
7,000 square miles, almost five million acres.

Large ranches were not unusual at the time; one of the most storied was
the XIT, created when the Texas legislature traded three million acres of
state land in the Panhandle to pay for construction of the new capitol in
Austin. The XIT differed from the JAL in important ways; investors, not
a working family, controlled the XIT and actually owned the land their
cattle grazed. Texas had retained all of its public lands when it joined the
United States, but the Territory of New Mexico had abundant federal
land available—anyone could use the public domain for free, limited only
by scarce water supplies or "prior possession." Unlike on the XIT, no
fences were originally built on the JAL. The Cowdens "did not begin
fencing until 1895, when they fenced over 300 sections," Cochran
recalled. "They did not own a bit of this land but later bought forty acres
of land around their wells." The frugal Cowdens saw no reason to buy
frontier land; in this the JAL was typical of other open-range ranches, as
Webb noted in *The Great Plains:* "As yet no ranchman owned any land or
grass; he merely owned the cattle and the camps. He did possess what was
recognized by his neighbors (but not by law) as range rights. This meant a
right to the water which he had appropriated and to the surrounding range.

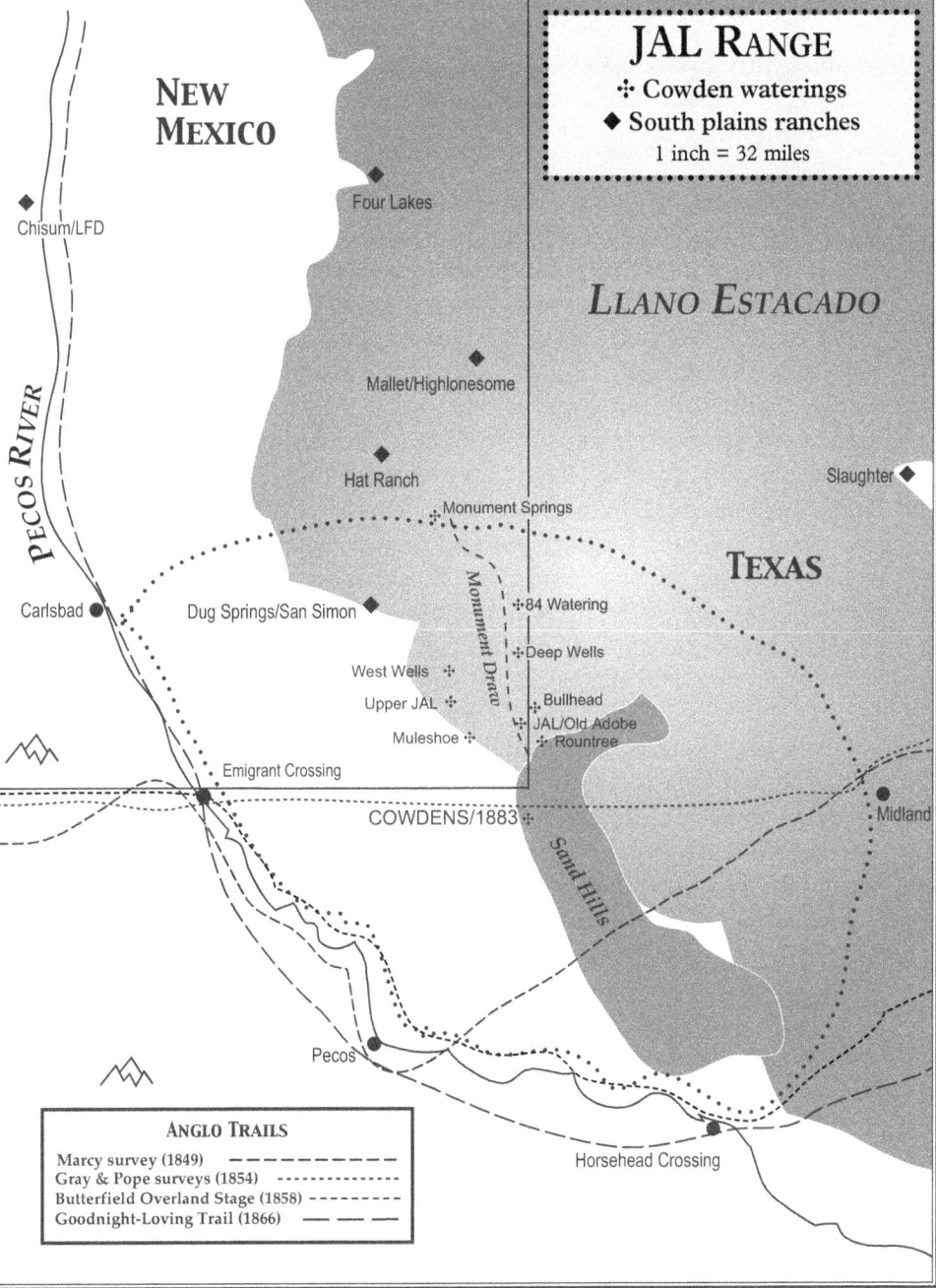

JAL RANGE
✢ Cowden waterings
◆ South plains ranches
1 inch = 32 miles

NEW MEXICO

LLANO ESTACADO

TEXAS

◆ Four Lakes

◆ Chisum/LFD

◆ Mallet/Highlonesome

◆ Hat Ranch

◆ Slaughter

PECOS RIVER

✢ Monument Springs

Carlsbad ●

◆ Dug Springs/San Simon

✢ 84 Watering

Monument Draw

✢ Deep Wells

West Wells ✢

Upper JAL ✢

✢ Bullhead

✢ JAL/Old Adobe

Muleshoe ✢

✢ Rountree

Emigrant Crossing

COWDENS/1883 ✢

● Midland

Sand Hills

Pecos ●

Horsehead Crossing ●

ANGLO TRAILS

Marcy survey (1849)   — — — — —
Gray & Pope surveys (1854) ·············
Butterfield Overland Stage (1858) ----------
Goodnight-Loving Trail (1866) — · — · —

Where water was scarce the control of it in any region gave control of all the land around it, for water was the *sine qua non* of the cattle country. . . . The ranchmen were careful to recognize that possession of water gave a man rights on the range. Moreover, it was not good form to try to crowd too much."

As the JAL herd rapidly increased, other ranchers were moving into the country. Droughts and overstocked ranches in Texas sent cattlemen looking for ranges up on the Llano Estacado, which had splendid grass if little surface water. "There never was any trouble in this country over range rights. This was due to the fact that this country was settled by old cowmen who respected the rights of one another," remembered Cochran. Droughts and blizzards actually promoted cooperation among cattlemen and revolutionary solutions to open-range problems; windmills and barbed-wire fences came into general use to address dry land and drifting cattle. "In the winter of 1884–5 there was the 'big drift.' Bad blizzards and snow caused it. I came out here in February, 1885, and went to the JALs and the whole world was working with cattle and trying to get them back to their ranges. Thousands and thousands drifted down to the Pecos from all over the Plains." Such conditions called for general roundups, where a number of ranches might send "outfits"—chuck wagons with a crew of cowboys and their strings of horses—to work the range. Smaller ranches simply sent one cowboy, an "outside man" who would join one of the larger outfits, a hand from the TAX or Hash Knife or 84 taking his meals with the JALs, working under the direction of their range boss, spending all spring helping to sort and brand drifted cattle before finally driving his stock back to their home range. James Hinkle recalled the period in *Early Days of a Cowboy on the Pecos:* "From 1885 and for about ten years thereafter it was just cattle, nothing more—only a fight to hold your own. The roundups started with the first green grass, usually in April, and continued until November and it was ride and ride hard all the time. We would send men out on the outside work and they would often be away for six months. They would work with the outside wagons and throw the cattle back towards the headquarters ranch."

After cooperating for several years to gather and sort scattered and mingled herds, ranchers north of the Cowdens—the Hat Ranch, Mallet, and LFD—eventually constructed fences to stop their cattle from drifting before winter storms. Fences that enclosed the public domain were illegal, but ranchers were far from any government that cared—the county seat of Lincoln was over two hundred miles away—and the barbed-wire "drift fences" that went up east and west across the Llano Estacado enclosed nothing. They marked ranges, if not legal boundaries, and the open range was still a fact of life. Imaginary lines in the sand were all that separated most cattle and cattlemen.

With enormous territory to cover, the Cowden brothers welcomed old friends and family who came out to join the JALs, including younger brothers Liddon, Rorie, and Eugene and their nephew Spence Jowell. Gene and Spence both arrived as adolescents, eleven and thirteen, to begin their lives as cowboys and cattlemen—a dream boys have dreamed from the days of the old Wild West to present times. In the new, spare, and sparsely settled country, the big Cowden family was a necessity at first. Cochran arrived, so close to the Cowdens in spirit and enterprise that he was called Uncle Walter. Sons of George Franklin Cowden also moved from Palo Pinto; G.F. Junior and Fred bought wells from their cousins in 1885, raising cattle first in the sandhills, then New Mexico, then back in Texas near Midland, which became a Cowden town. "There was a large family of these people," noted Susie Noble in "History of Midland County." "When we came here, an uncle of ours said never talk about anyone here for if they weren't a Cowden, they would be kin to them and it might start trouble." Jeff Cowden, who had roped the train back in Palo Pinto, moved to Dallas rather than stay in Midland, where he figured his only choice would be to "marry a cousin or a jackrabbit."

As the JAL herd accumulated and scattered over the country, growing to seven thousand head by 1886, more hands were hired to ride the range, and the Cowdens began to make the transition from "cowboys"—those in the employ of others—to "cowmen," who paid the wages. "You know the difference between a cowboy and a cowman?" Guy Cowden once asked. "A cowman owes the bank money." Cowboys enjoyed their relative freedom from responsibility, but most aspired to become ranchers themselves, a dream difficult to realize in part because the land and weather dictated a large scale of operations. Individuals couldn't make a go of it; the 160 or 320 acres a man might homestead were insufficient to support enough cattle to make a living, and, without irrigation, farming was impossible. A common practice evolved where cowboys would homestead land with the understanding they'd sell it later to the outfit that paid their wages; most often such homesteads were at actual or potential locations of wells, helping big outfits like the JAL maintain control of vast ranges. JAL cowboys could also run their own cattle with those of the outfit, as Cochran noted: "I had shipped what few cattle I had from Palo Pinto in 1884, and turned them loose on the Cowden Ranch in the Sand Hills about forty miles north of Monahans. When I went to gather them in 1886, I found them scattered from Horsehead Crossing on the Pecos to Roswell, New Mexico. I had just enough cattle to have a working interest in all the big round-ups that were made on the Pecos. The Cowdens always used me as their man 'Friday,' so it always fell to my lot to have to make all the big round-ups on the Pecos. When I sold out to the Cowdens in 1894, I had picked up a good little bunch of cattle."

From their arrival in 1883 until after the turn of the century, the Cowdens built the JAL into an impressive and profitable open-range ranch—by 1895 the Cowden Cattle Company was estimated to own forty thousand head of cattle. May Price Mosley recalled, "The San Simons—as did the Cowden Bros JAL Outfit—continued to increase their range by developing new waterings, and by buying out the few early settlers within their grazing range. . . . There was at this time a strong tendency toward larger ranches in this section of the plains rather than smaller. From 1890 to well past the turn of the century was the era of the big ranches here—the heyday of the free grass, open range ranching; and there was little further change in ownership until after 1900."

Soon, however, the JALs and other large open-range ranches would face increasing pressure from settlers, "nesters" filing claims for 160 acres of public land under the Homestead Act. Designed originally to relieve pressure on overpopulated eastern cities and to blunt the expansion of slaveholding interests, the Homestead Act was passed in 1862, only after Southern states had seceded from the Union. It now threatened to cut up western cattle ranges, unless ranchers bought out or drove off homesteaders. Free and open rangeland was coming to an end. After the turn of the century, with improved well drilling and windmills, more and more homesteaders were encouraged to try their hands on Llano Estacado and other rangeland. "By 1905 homesteading settlers were dotting and fringing all the ranges of the old ranches," remembered Mosley. "Soon they became an actual menace to the ranching industry. 1910 saw the old ranges cut every way by settlers, small ranchers, and settlements." When New Mexico became a state in 1912, lease laws for public lands further diminished free ranges, and the Cowdens began to dissolve their long, successful family enterprise. John M. Cowden bought out his brothers' interests in New Mexico, and the Cowden brothers began to seek out individual ranches there and in West Texas.

The people, land, and labor remained, but a legendary era was over.

Tumbleweeds

May 2001

April 24, 2001—Another flight to New Mexico for another spring roundup and branding, another round of research. Circles everywhere: cycles of seasons, recurring labors, time turning and returning. I now realize I'm the "outside man" for Cowdens past and present. Rather than gather stray cattle, I assemble their stories, scattered far from the home range. Of course, Sam is happy to have me punching cows rather than a keyboard. Whatever—I ride for the brand.

Before heading to the Cowden Ranch, I'm driving down through the old JAL Range. I'll make a big swing through the territory, visiting Bill Cowden on his ranch in the Davis Mountains and Jon Means on the Moon Ranch— two more of the cousins who seem to be everywhere, like jackrabbits in the skinny shade of mesquites. Along the way, I'll stop in Roswell, where my grandfather went to school at New Mexico Military Institute. And spend a couple of days in Midland, searching for stories false and true about the Cowdens.

My main quest is to locate Old Adobe, working headquarters of the JAL from 1886, when it was built somewhere on Monument Draw. I've got a lead on the site from Brian Norwood, the artist who conceived, built, and installed the Jal cowboy sculpture I saw last year. I want to walk the "hallowed ground" of the JALs, although I bet no cowboy who ever spent a blazing summer riding over that range thought of it in such terms. I can hear the response: "Hallowed ground? Hell it was!"

April 28, 2001—Roswell, New Mexico. Ground zero for UFOs in the USA. Downtown, you can visit the International UFO Museum and Research Center or check out the 1947 crash—they say—of a flying saucer on a ranch northwest of town. Or sip coffee at the Alien Caffeine Espresso Bar.

Or make more scientific inquiries into the heavens at the Goddard Plane-tarium, maybe at their Kids' Space Camp each summer. The planetarium is named for Robert Goddard, a pioneer rocket scientist who worked here for the U.S. Navy—how weird is that, the navy in landlocked Roswell?—but hailed from Worcester, Massachusetts, where the Cowdens first settled in the eighteenth century.

It makes sense that folks in Roswell look to the skies, because the earth here is not entirely rewarding. It's country most folks hurry through, although the Pecos provides irrigation for local farms, and there's a greenbelt along the river. The Pecos and its tributaries prompted John Chisum to establish his ranch here in 1873. Chisum—often confused with Jesse Chisholm of the Texas-Kansas cattle trail—was remotely involved in the Lincoln County War that brought Billy the Kid and Pat Garrett to prominence, but he's best known as the cattle king of the Pecos. Cattle with his jinglebob earmarks grazed up and down the river for 150 miles. To Chisum's east, Comanches roamed the Llano Estacado, and to his west, Mescalero Apaches swept out of the Guadalupe and Sacramento mountains on raids. Early drovers like Loving and Goodnight often crossed their herds to whichever side of the Pecos promised the least threat. It didn't always help; Indian raiders mortally wounded Loving near here, and he died up in Fort Sumner, where Pat Garrett later snuffed Billy the Kid. Do you suppose those aliens crashed their spacecraft while checking out old Wild West sites?

But I'm not in Roswell for flying saucers or gunslingers. I'm here because NMMI is here. Founded in 1891, New Mexico Military Institute now offers high school and junior college to a thousand cadets, among them my cousin's son. I watched an NMMI baseball game this afternoon with Jon and Jackie Means—players sending fly balls up into the ever-present plains wind, which routinely carried them out of the park—and I'll visit them tomorrow on their ranch down in the Trans-Pecos country of Texas.

NMMI has changed over the years; girls now make up about 20 percent of the corps. When I wondered why girls would attend military school, which echoes of boys with behavioral problems, Jon and Jackie pointed out that NMMI is a public institution and it's possible to get a good education at little expense. Many students come from out of state as well; Guy and Luke Kellogg both attended in the 1980s. Guy Cowden was a cadet here in 1916—a mismatch of man and mission if there ever was one. I've seen pictures of him in uniform and read the letters he sent home to Midland, thanking his mother for care packages, but it's hard to place him in such disciplined surroundings. But the boy with behavioral problems I picture at the military school was in fact Big Guy.

Back East, military school is a last, rather than best, resort for educating children, but here it's a reasonable choice for ranching families far from town and for parents facing lousy school systems. Rooster Cowden hated his time in the military during World War II—another New Mexico navy man—but he gave Sam the option to attend NMMI. Like Rooster and Guy before him, Sam always heard the call of ranching and cared little for the schoolhouse bell; he chose to stay in Santa Rosa.

NMMI has a handsome campus, yellow brick buildings with crenellated towers, like a Foreign Legion fortress. There's a general, a commandant, and the usual military procedures and paraphernalia. This motel is named for the campus "Sally Port," a portal under a four-story tower attached to the old barracks; from there cadets would "sally" forth to various classes, drills, and activities. NMMI once had an accomplished polo team—Jon Means's grandfather and father both rode here—but no more. However much a cowboy loves his pony, I'd bet the cadets are happy to have girls instead.

Before any coeducation here, Sam's sisters went to boarding school in El Paso, and other ranch kids now go to Dallas or even more remote schools, looking for "real" civilization and culture. It's commonplace that rural folk often discount or don't recognize their own culture, and there is great value in exposure to other ways of life, but I don't dismiss this place. Not that I could hack it. Too much Big Guy in me for so much regimentation; military school would make me a behavior problem.

※ ※ ※

April 29, 2001—I'm in the guest house on the Moon Ranch, eighty-eight thousand acres west of the Davis Mountains in Texas, where my cousin Jon Means and his wife, Jackie, live. The Meanses have been prominent Texas ranchers for generations; when my grandfather's older sister, "Sissy" Cowden, married "Bug" Means, the union had all the qualities of a state-arranged marriage of ranching royalty. But then, in such sparsely settled country, who was available to men or women but their counterparts?

Before coming here, I stopped to visit with Bill and Mary Cowden at their ranch on the edge of the Davis Mountains. Bill's grandfather was the eldest of the brothers who started the JAL—William Henry Cowden, whose portrait hung in the Hall of Cattle Kings of the Texas Centennial Exposition with this biography:

44. WILLIAM HENRY COWDEN was born October 6, 1853, in Shelby County, Texas. His parents were William Hamby and Caroline Liddon Cowden. On October 20, 1880, he married Mary Salvage, daughter of Benjamin Gilbert and Harriet Atwood Liddon Salvage. In 1882 he and his brothers, George and John M., bought the "J A L" brand and about one hundred head of cattle from Alonzo Edwards. Mr. Edwards had married the widowed mother of Mrs. Wm. Henry Cowden. The Cowden brothers established the "J A L" Ranch in Southern New Mexico and adjoining lands in Texas. They operated this ranch twenty-five years. The firm in time included the younger Cowden brothers, Liddon, Rorie and Eugene, as well as the three older ones. In 1912 John M. Cowden bought the interest of all his brothers. Wm. Henry Cowden then acquired ranch land in Crane County and later in Frio County. He was president of the First National Bank of Midland from 1890 to 1926. He moved to San Antonio in 1926 and died there July 20, 1933.

"Uncle Billy," as most knew him in Midland, was an important figure in the Cowden ascension to West Texas ranching prominence, in no small part because of his banking connection. When the First National Bank of Midland failed in 1983, it was the largest independent bank in Texas, rising first with the fortunes of West Texas cattlemen, then with the Permian Basin oil business. Even though Uncle Billy sold his interest in the JALs to his brothers in 1892, he clearly led the move west from Palo Pinto, and I hoped Bill Cowden had stories about the JALs direct from his grandfather.

Before we sat down with a tape recorder, Bill and I took a drive up Gomez Peak, northernmost of the Davis Mountains and the physical backbone of Bill's ranch. It's been extremely dry the past few years, so Bill isn't running any cattle right now and Gomez Peak provides income from two radio towers up near the crest. With an unobstructed view for a hundred miles west and east, from Van Horn to Fort Stockton, one of the towers provides cell phone service to passing motorists and the few residents along the I-10 corridor. Bill had the other tower erected and leases space on it to commercial customers. We also saw the hunting camp Bill leases out; the Davis Mountains have a healthy population of deer and other game—natural resources still sustain the Cowden family.

While coming off the mountain, we watched a thunderstorm move in from the west, always a beautiful vision for ranchers in dry country. By the time we reached the house, it was blowing hard and spitting a little rain. Just after we sat down with my tape recorder, the power went out, as though the storm were attempting to extinguish any record that it made an appearance—an ephemeral passage through a durable landscape. Thanks to Eveready, I was still able to record some Cowden history from Bill's perspective, most of it familiar material.

Then I made the forty-mile drive here, Moon Ranch headquarters with its lovely house facing the Davis Mountains. Thirty miles south is the Rio Grande and the Mexican border; the Trans-Pecos country—that part of Texas west of the Pecos River—shows profound Hispanic influence, always has. Much of the rest of West Texas—the South Plains and the Panhandle—wasn't permanently settled until Anglo cattlemen pushed west after the Civil War. Now, of course, there's great pressure on the entire southwestern United States from Mexico and Latin America; combined with native Hispanics, legal and illegal immigration has made Anglos here a minority culture, although they retain considerable wealth, power, and influence. Jon told me that a family from Mexico joined them in buying and dividing a ranch adjoining him, demonstrating that there's more to Mexican immigration than a poor, manual labor pool.

Jon and Jackie—who went to Mount Holyoke College in Massachusetts, where I taught for years—had lots to say this evening about ranching life. Like Sam and Kathy, they are concerned for their livelihood and their family. In short, for their whole way of life. "Not long ago I could have named nine contiguous ranches out here," Jon said, "that didn't have any cattle on them. Absentee owners bought these huge ranches—sixty or eighty thousand acres, a six- or ten-million-dollar investment—just for fun. No commercial hunting, nothing. Their family's private hunting ranch. They'd come out maybe five days a year." As a cattleman, Jon is puzzled by grazing land going to waste, and the absence of other active ranches limits opportunities for the Means children, which is what worries Sam and Kathy—no neighbors, little exposure to other kids. So it's off to boarding school, and then how are you going to keep them down on the farm? The flight of ranching families sounds like a southwestern domino theory.

One neighbor actively ranching is actor Tommy Lee Jones. Born in Texas, Tommy Lee had a scrappy childhood and played football in Midland before heading off to boarding school in Dallas, then Harvard, then Hollywood—a real-life Jett Rink, the rags-to-riches character James Dean played in *Giant*. Except that Tommy Lee isn't flashing his good fortune in everyone's face. "He's trying to do it right," says Jon. "Really ranching. Taking care of his fences, improving his waterings. Not just throwing money into a fancy house." Jon shows the modest streak typical of Cowdens and Meanses and other longtime ranching families—you'd never know they were "cattle kings." I remember driving with my aunt Olive Cowden around the old graveyard in Midland, where we visited my great-grandparents' simple stones, then came across a new, ostentatious tomb. "My word," Olive marveled. "Can you imagine that?" Nothing like life on the range to keep you down to earth, even when heaven calls.

April 30, 2001—Midland, Texas. Hometown of George W. and Laura Bush. So they say—both the Bushes, looking for political capital in the "family values" of a conservative Texas city, and the rest of Midland, looking for any distinction it can muster. Midland has always served as corporate and financial center of the Permian Basin; twenty miles southwest, Odessa is home to blue-collar, oil-field services. Where Midland has high-rise bank buildings, Odessa has gritty rig and pipeline supply centers. Rivalry and ill will between the two cities has always been intense.

Yet oil rules all. The Permian Basin phone book carries twenty-two yellow pages of listings under Oil and Gas Exploration and Development, Oil and Gas Producers, Oil Field Construction, Oil Field Equipment, Oil Field Service, Oil Field Supplies, Oil Land Leases, Oil Operators, Oil Well Services, and such. You can visit the Petroleum Museum, the museum's Archives Center and Library, and the Petroleum Hall of Fame in Midland; I will, to see how Big Oil blew in and changed the face of the landscape.

I get ahead of myself. This morning Jon and I drove around the Moon Ranch, looking at land and cattle. He runs Angus cattle now, a fundamental shift from the Herefords that used to dominate West Texas. When cattlemen began improving the stock of Longhorns that built the Cattle Kingdom, Herefords became the breed of choice for much of the twentieth century—certainly among the Cowdens. Now the move is toward Black Angus cattle. "Up in Kansas, where we get our bulls," Jon told me, "a lot of big Hereford ranches are buying black bulls. The Scharbauers, the Four Sixes. Ten years ago you couldn't have gotten any of those guys to even fly over that herd." Jon used to produce black baldies but now has straight Angus stock, so he can sell calves at a premium as replacement heifers or bulls. "Before, whatever we sold was by the pound," Jon said. "Selling by the head made more sense for us economically. On our lease country—some of it pretty marginal country—we still have some crossbred cows."

That marginal lease country includes the Dollarhide Ranch west of Midland, situated on cusp of the Llano Estacado and the Chihuahuan

Desert, with sandhill formations limiting grazing and requiring scrappy cattle. Although farther south in the Chihuahuan Desert formation, the Moon Ranch benefits from the Davis Mountains' tendency to produce more rain and better grazing. You still know it's desert, peppered with creosote bush, Spanish dagger and other yuccas, and cholla, ocotillo, and prickly pear cactus. "Our worst deal," Jon said, "is a world of greasewood. It's good for nothing." In the flats of the Moon Ranch, tobosa grass produces abundant, if coarse, grazing; grama grasses predominate elsewhere. Like Sam's place, the Moon Ranch has adequate shelter but not much rough, unproductive country. And like Sam, Jon is careful about stocking rates. "We ought to run fifteen hundred cows here," he said. "Obviously we don't have that now. A lot of times, twelve hundred cows are good. Probably, conservatively speaking, a cow to fifty acres." I note that Jon uses "conservative" as a physical, not political, term. Despite the frequent opposition between ranchers and environmentalists in the West, both aim to conserve natural resources. If not, a cattleman can quickly ruin his ranchland and find himself driving a truck. Stupid or greedy ranchers will overgraze rangeland, often when leased from either private or public sources.

The route between the Moon Ranch and Midland took me through Pecos and then through JAL rangeland, sorry-looking stretches of mesquite, creosote bush, and greasewood. At Monahans the sandhills began, with their hot, dry invitation to hurry on. I did, through oil and gas fields as I passed almost imperceptibly up onto the southern reaches of the Llano Estacado; no distinctive caprock marks this edge of the plain. I scooted past Odessa and on into Midland, shimmering in the heat.

This afternoon I went to Fairview Cemetery, but no one was in the office. I located William Hamby Cowden's gravesite on the plot map and eventually found his grave. He's buried beside Kittie Liddon Moore Cowden; next to them are W. C. Cochran and his wife, also a Cowden girl. Across the path is John M. Cowden, the whole kit and caboodle hanging together in death as in life.

Now, having taken vows of poverty to write this book and live this life, here I am in the cheap but clean Metro Inn. In the parking lot underneath the windows and doors of the rooms are old Econolines, diesel pickups of oil-field workers, battered sedans of old men, trucks with Coahuila or Chihuahua plates, and my Toyota Camry. It's not, despite appearances, a bad place—no loud nights of arguments or fistfights. Just a place to return to after my days in the library and courthouse, a place to plug in my laptop and recharge my batteries.

✳ ✳ ✳

May 1, 2001—Back after a full day of research at the courthouse and the Haley Library. Tonight I'll have supper with Bill Jowell and Spence Collins, both distant cousins, part of the great Cowden diaspora of the nineteenth century.

At the Midland County Courthouse, I had to stash my pocketknife in a locker and pass through a metal detector to enter. When the Cowdens came to register their marks and brands a century back, did they have to turn over their handguns and calf knives? Have times changed all that much since 1892? On July 7 of that year, the Cowdens—perhaps ten of them by my count—poured into county clerk O. B. Holt's office. The first brand they registered, in Book 1, Marks and Brands Record, Midland County, was

Cowden Bros. & Cochran. Midland Tex.

Mark: ⌣⊃⌣⊋

Brand: J A L on Left Side. Kept up. h. s. sh. [hip, side, shoulder].

⌣⊃⌣⊋ I H X on Right side & C on right hip.

+ - C right side. ⌣⊃⌣⊋

C right hip. T I X Right Side ⌣⊃⌣⊋

That same day, in succession, Holt also registered all of the following:

W. H. Cowden Sr. Midland Texas

Mark: ⌣⊃⌣⊋

Brand: C right hip   C right shoulder

J. M. Cowden & W. C. Cochran. Midland Texas

Mark: ⌣⊃⌣⊋ Swallow fork in right & under slope left ears.

Brand: <u>F C</u> on both sides Kept up

W. F & G. F. Cowden. & J. S. Strawn Midland Texas

Mark: ⌣◁⊂⌣ Grub the left ear

Brand: 2̂ on both Sides.

W. F. Cowden. Midland Texas

Mark: ⌣◁⌒⊂⌣ crop & underbit left. & grub the right

Brand: F E D on left side   C on left ~~shoulder~~ hip

J. T. & E. M. Cowden. Midland Texas

Mark: ⌣◁⊂⊃⌣ crop & split left & split right ears

Brand: F I X on left side   C on left hip

G. F. Cowden. Midland Texas

Mark: ⌣◁⊂⌣ Grub the left ear

Brand: 2 on each Shoulder

Liddon & R. W. Cowden. Midland Texas

Mark: ⌣◁⊂⌣

Brand: A L on left side   C on left hip

Although most of their cattle ran west of Midland and primarily in New Mexico, the Cowdens needed to register their brands in any county where their herds ranged. Cattle didn't know, nor did cowboys much of the time, whether they were in one state, one county, or the next. I'll also check for marks and brands in Ector, Winkler, Crane, and Andrews counties, Texas, as well as in New Mexico, where brands are registered statewide.

The next time the Cowdens appear in the Midland County records is September 10, 1895, when they put O. B. Holt through another complicated round of marks and brands, as the brothers apparently refined the complex, shifting partnerships among themselves:

Geo. E. and J. M. Cowden Midland Tex

Mark: ⌣⌣

Brand: J A L on left side   J A L left side \ on left leg Kept up

⌣⌣ I H X on right side   C on right hip right hip + right side –

C right shoulder ⌣⌣

⌣⌣ C right hip   T I X right side ⌣⌣ A L N left side

No cowman wanted to employ too many branding irons, so permutations and combinations of basic marks and brands could get confusing. But if your life was working cattle on the open range, you became adept at recognizing what belonged to whom, especially when dealing with family. With a family as large as the Cowden clan, the tally man at a roundup had to be particularly sharp, as cowboys dragging calves to the fire called out the various earmarks and brands of their mothers. "J-A-L! Crop and notch the left, swallowfork and underbit the right!" Or "A-L left side, C left hip! Crop and slash the left, underbit the right!" The wrong mark called out, a brand burned wrong, could cost someone a calf that year, and all her progeny in the future. Sam once told me that our great-grandfather, Gene Cowden, was tally man for the JALs; if so, he must have been competent and trustworthy.

After the courthouse run, I headed for the Haley Library, where I met and interviewed Spence Collins, whose grandfather was Spence Jowell, who came out from Palo Pinto to the JALs in 1886 with Gene Cowden—his uncle, although Jowell was older by a couple of years. Got that? Confusing, but that's what happens when you have twelve children over the course of twenty-five years, then sixty-five grandchildren, and so on. I hardly try to keep it straight. I just call everyone cousin.

Spence Collins retired from the oil business after working in South and Central America for most of his career. He went to the Texas School of Mines—later the University of Texas, El Paso—to escape ranching life. "We had a string of two-year-old horses," Spence said. "And I had one of these colts roll over on me. Didn't hurt him or break my bones, but I'm telling you, for a week I couldn't move. And that's when I decided: I'm going to do what my aunt told me to do—I'm going to school." He hasn't worn boots or jeans since he left. "It drives Billy crazy that I don't own a pair of boots," said Spence, dressed like a well-to-do businessman, while Bill Jowell always sports jeans and boots. "But they only remind me of poverty. It was hand-to-mouth and day-by-day."

Spence Jowell had been a profound influence, "a true patriarch," Spence said. "He was a very kind person. I don't believe I ever heard him say anything bad about anybody. Very honest. High integrity. And a hard, hard worker. You got up in the morning and did all those women's chores before daylight, so you'd be at the back side of the pasture waiting on daylight to come. So you could go to work. And then the things that you had to do when you came in at night—chop wood, milk the cows, clean up around— you did all that. That was just the way of life."

I stayed at the library, where archivist Jim Bradshaw had the Spence Jowell and Bob Beverly collections ready for me, both of them with materials on the JALs. My research hadn't produced a bonanza of family letters or journals; only the Walter Cochran memoirs provided a clear sense of Cowden and JAL history. So any ranching records, however spotty, were important. Like Cochran, Spence Jowell had worked with his Cowden relatives on the JALs and eventually became range boss; Bob Beverly became ranch manager in 1912, when the JAL was closing out operations. Between the two collections, I found photographs, tally books, and interviews conducted by J. Evetts Haley. I couldn't digest everything, but one thing became immediately clear—the nature of open-range ranching, as shown by records kept on the range. One day's tally, on November 12, 1902, lists strays branded

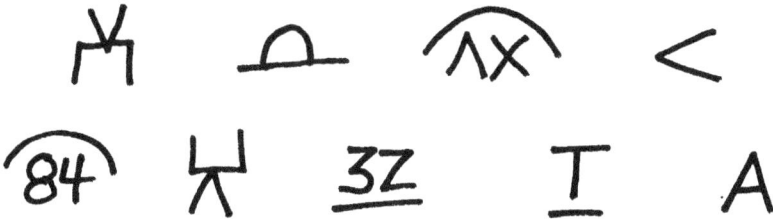

The next entry, November 13, near "Upper Cooper & Moon," records a dozen brands:

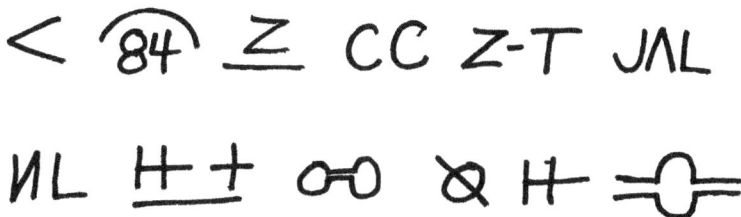

Even after the turn of the century as windmills, fences, and legislation were bringing an end to free grass and wandering cattle, the open range was still a fact of life.

Note cards by J. Evetts Haley quote Bob Beverly: "John M Cowden branded H; Billy Cowden had 2 or 3 diff. Brands; Walter Cochran came in with the Muleshoes, & located at site of town of JAL. Doc Cowden had 2's. . . . Old man Bill & other Cowdens had long hard fight—old man Bill sd., he figured he had to feed them. Never sd. Anything but just kept on raising cows. After JALs bought 84s—I guess JALs had most cattle. Gene & Rory bought 84 in 1902, & then they proved Co.—When I went out & took chg. as foreman in 1912, they had 32 diff. Brands. I went to running straight JAL brand. Were 10 of Cowdens in Co. & everyone had diff. Brands. I left shipping office in Nov. 1912, & left for JALs next day. Gene a fine man, & best cowman of bunch. . . . I told them to all get together & pick out one man to represent them; I wouldn't take orders from anybody else. They picked Gene."

My little heart swelled with pride at that: my great-grandfather the "best cowman of [the] bunch" of Cowdens who built the great JAL Ranch. I've also seen enough conflicting accounts to know that any one particular source can be wrong, and Beverly's observation is opinion, of course. But he wasn't family and had no reason for deliberate misstatement. And it makes sense, since Gene came out to the JALs at eleven or twelve years old and worked for years with his experienced older brothers. Here was a man who knew cattle ranching inside out. So I can take legitimate pride. Gene may have been the youngest child but evidently grew into the name we always knew him by: Big Daddy.

Another revelation from the Beverly collection was the "McKenzie War," which involved the Cowdens. "General" John McKenzie—by his own admission not an actual officer, but "general of a band of Texas cow thieves"—ran cattle on free range north of Midland, until Texas began selling its unappropriated state land. Nelson Morris, a meat packer from Chicago, bought a large tract in 1883, forcing McKenzie's outfit farther west, much as European and American settlers had crowded various Indian tribes off their historic lands and consequently into conflict with tribes to the west. Beverly noted that McKenzie moved "the Dumb Bells out here at the draw and that was his headquarters. . . . George Cowden had started the JHS and Bill Cowden for the JAL—and McKenzie come in right between Billie and George and dug him a well on the draw there."

Aside from infringing on the Cowdens' range, McKenzie and his cowboys evidently built up their herd by unsavory practices such as killing cows so that their unbranded calves might be burned with the Dumbbell brand

and no one would be the wiser. At least no one could prove anything. According to Beverly,

> What started the whole thing was, they got so rank down here and was killing those JAL cows and everybody else's for the calves, so they decided that there wasn't nobody going to work their [Dumbbell] range, and Cowdens couldn't work their range. They was afraid to.
>
> All this Southern New Mexico was settled with a bunch of thieves. All of them fellows that come out with McKenzie, they left a lot of old sprouts. That was the job given me when—come from Midland to southeastern New Mexico in 1912—to tell these people that law had arrived. The Cowden Cattle Company and the Association sent me. . . . They got afraid to go out and grease their windmills. It was the worst thing I ever saw in the way of cattle stealing.
>
> Old Bill, John M., and the George Cowden bunch never did get into that. They were all clean men. . . . That bunch of Cowdens were mighty good, straight men.

I was relieved to read that the Cowdens were not cattle thieves—my family could release a collectively held breath.

When McKenzie wouldn't permit other outfits to work his range and gather their stray cattle, the Cowdens and the cattlemen's association sent in Beverly and others to reclaim their stock. Wagons and outfits from other ranches were there for the same purpose, everyone armed. Beverly quoted one cowboy's report to McKenzie, "'General, we are all going to get killed! . . . There is an old man over there with a shotgun so long he has to get on a windmill to put shot down the barrel. He is going to kill us all.'"

I sat in the quiet, cool Haley Library—absorbed, edgy, fascinated. Who would draw first? Who would draw fastest? Who would live and die? 1912—it was the twentieth century. New Mexico had become the forty-seventh state on January 6—which helped prompt the roundup, since the state now could lease public lands to private individuals, who might then fence it—but it was barely settled. The events echoed battles with Indians on the frontier and standoffs between cattlemen and railroad crews. As Beverly reported,

> MacKenzie went to see the Association man and said, "Do you want to work this range in six shooter smoke or in peace?" He said, "It don't make any difference." The General said, "Well, we will work it in peace. We will stack all the guns out here and let no man have a gun."
> They went out and stacked their guns and set a bunch guarding those guns. He was pretty clean cut and wouldn't let anybody have a gun. They went to work and worked the range and rounded up the whole country. When it was all over, Cowden was (still) afraid to claim his cattle and lots of other people [were too]. The Hash Knives got theirs. And that passed off as the MacKenzie War.

As in so many confrontations, this was a game of chicken—no feathers flew, no bullets whistled, one side backed down—but dangerous nonetheless. "It was a rather difficult thing to handle," Beverly observed. "[The Cowdens] had a world of stuff, and the country was beginning to settle up, and you had to get the stuff out or lose it, that was all. I gathered the Cowdens' stuff and wound that up and turned it over to John M."

A good day of research, and tonight I'll hear more from Bill Jowell and Spence Collins. Tomorrow I make another pilgrimage to the JAL Range.

May 2, 2001—Last night I heard tales of the generation that followed the JAL era and faced a series of profound changes: final days of open range, World War I, Prohibition, the Great Depression, the Dust Bowl, and the West Texas oil boom, when lifelong ranchers suddenly struck it rich. Important developments, but most writing about cattle ranching focuses on trail driving and open-range days, a more romantic time. The arrival of the twentieth century seemed to bring ranching to an end, not a beginning—not until oil resources were developed did ranching again see good times.

Midland was, and still is, the center of the Permian Basin oil business, and at the heart of Midland was the First National Bank and the Scharbauer Hotel across the street. Bill Jowell says that Guy Cowden, who spent his share of hard-drinking days, would hire the "colored shoeshine boy" at the Scharbauer Hotel as his driver. Van Dyke—Bill couldn't recall his first name, because the odd last name stuck to the exclusion of everything else—would chauffeur "Mr. Cowden" all over Midland and West Texas and New Mexico, so my grandfather could drink to his heart's delight or despair. Guy eventually quit drinking altogether—only once did I see him sip a little wine. But Sam tells the story of a roundup on the ranch, at Sabino Springs, a difficult gather completely wrecked when Big Guy's car came racing up in a cloud of dust, scattering all the cattle. It was Van Dyke at the wheel, and Big Guy beside him with a bottle.

Guy Cowden was not alone in his mischief. Many cattlemen—with the world by the tail suddenly, oil royalty checks coming in, ranches and herds growing—pulled stunts legendary in Midland. There was Buck York in his hotel room in Fort Worth, calling a local Cadillac dealer to order a new Sedan DeVille and asking the dealer to "deliver it and just charge it to my room." "Excuse me, Mr. York," said the flustered sales manager on the telephone. "This is a little out of the ordinary. Who did you say you were with?" Rather than provide the name of some familiar and prosperous company like Humble Oil or Texaco, Buck York turned from the phone and was heard to inquire, "Honey, what did you say your name was?" Buck got his Cadillac, however, and paid his hotel bill with its charge for a Sedan DeVille in cash.

A look of sly admiration crossed Bill Jowell's face when he related that story. Another antic involved a train "borrowed" in Mexico, from Juárez or some other border town, which raucous cowboys—not my grandfather— drove deep into the heart of Chihuahua or Sonora before they sobered up. Lucky they weren't shot or imprisoned by outraged Mexicans—they probably talked themselves out of trouble with dollars.

Yet another joyride featured a private plane. Although they'd all flown often with private pilots, none of the rowdy ranchers were themselves licensed, and none had ever flown drunk, the apparent stimulation for their trip. They managed to take off, wobbly, dipping, saved from disaster by the grace of God, and flew around awhile before coming in for their landing. That landing was marked not only by their amateur, alcoholic efforts but also by a herd of javelinas that happened to scurry out onto whatever runway they were using—likely some ranch road—just as they were coming in. "Oh shit, boys, it's a herd of javelinas!" The plane hit the wild pigs, flipped over, yet avoided total disaster. The cowmen stumbled away from the wreckage intact.

Also legendary was the morning that some of the Midland crowd decided they could not do without coffee from an open campfire. Trouble was, they weren't out on the range but in a hotel room—I've heard it told about both Fort Worth and El Paso. The story goes that the hotel burned down.

Apocryphal? Factual? Every culture has myths and legends, and cowboy culture may have more than most. Pecos Bill—bumped from a jostling wagon full of pioneer children, raised by coyotes, outrageous and irrepressible—is a variation on trail-driving cowboys who would tear up Dodge City at the end of a dry two-month drive. Guy Cowden, Buck York, Foy Proctor, George Glass, Holt Jowell, and others were variations on Pecos Bill and their own fathers. What's the moral of the story? Beyond pointing an easy finger at alcohol, there is no moral. The stories don't aim to teach good and bad behavior; they are entertaining. What a character Guy Cowden was! When he bought the old toothless circus lion, hauled it out to the ranch, chained it to a telephone pole, figuring to run it with his hunting dogs—what a character! How unlike anyone else and how absolutely himself! "Mr. Cowden," said his son-in-law Walter Kellogg, "I'm not sure that's a good idea." And Big Guy was not even drinking.

We admire these tales for their excess, but most people want someone else to go through the demands of debauched living, its romantic self-destruction. Let Buck York or Guy Cowden play the wild man, with his driver Van Dyke and his Cadillac and his long drunks and crying wife and embarrassed children. His life makes for a good story, if hard to live through. "God save us from interesting times" goes the plea. Guy Cowden eventually saved himself and his family by giving up the bottle.

Time to take off for Jal. Out to that spare country that put such a thirst in so many men.

A big day, a good day. Drove this morning to Jal, where I met Brian Norwood at the D&N Restaurant. We were joined for breakfast by a couple of Jal residents: Carroll Leavell, state senator from Jal, and Jack Hedgpeth, whom I thank for my meal. Brian has become a local celebrity and, with his cowboy sculpture, has probably done as much for Jal as anyone in town. Brian hopes to install cowboy sculptures the whole length of Lea County, from Tatum all the way south to the Texas line—between Lovington and Hobbs and Jal and at the Muleshoe watering. It's a wonderful, ambitious plan, worthy of support from private and public sources. But do they exist, on such a scale?

After breakfast, Brian and I crossed the street and walked to the Muleshoe watering, a few undeveloped acres in the middle of town. "Undeveloped" is misleading—Jal is both small and sprawling, lots of empty space here and there. Muleshoe was constructed and used by Walter Cochran after he came out to join the JALs in 1885. He eventually married one of the

Cowden girls and was called Uncle Walter with more justification than usual—almost everyone back then was familiar enough to be called Uncle or Aunt, whether or not they were. Muleshoe watering took its name from the Cochran brand brought from Palo Pinto, and the town of Jal eventually grew up at the site, taking its name from the Cowden brand. Now a thicket of mesquite, scrub oak, and prickly pear around evident remains of a watering hole, Muleshoe is a Staked Plains oasis. Anywhere in this country you find substantial vegetation, you can bet there was or is water close to the surface. Refreshing, providing green shade from the sun that everywhere else baked the weathered buildings of Jal and open landscape beyond, its presence must have stimulated in Brian an appreciation of those pioneering cattlemen. We share a debt to them.

Afterwards, Brian and I met Gary Blocker, whose family had a ranch near Old Adobe. Once mayor of Jal, Gary drove us out Highway 128 toward the Texas line, filling me in on local history. I've got it all on tape— early ranching and later homesteads, the search for water and oil, details of life in a demanding land as cattlemen, homesteaders, and oilmen have come and gone over the years.

All the while, we were making our way toward the site of Old Adobe, some seven miles northeast of town on Monument Draw, not far from the Texas line. We turned north off the highway on an oil-field road, same one I explored last year—Dollarhide Road, where signs announced Union Oil's West Dollarhide Field. Along the way, Gary told me about land use, improvements, and the Old Adobe site: "Years back, the [Environmental Protection Agency's] Great Plains Program would pay up to a certain amount for the cost of mesquite eradication. Like most government programs, it ran until someone said this is a boondoggle, and they cut it. . . . We grubbed out this draw with bulldozers. You see that some of it is coming back, some of it's not. Some of it's also been sprayed. . . . Well, when Mr. Cooper was bulldozing mesquite on his place, he just didn't stop. Bulldozed down the Old Adobe and anything else in his way." I could only shake my head at the loss. Gary told me about another occasion where a dozer driver destroyed the remains of an old dairy that homesteaders had attempted to operate. "We asked him, when you saw those bottles, why didn't you stop? He said, 'I don't know. Didn't know you wanted them bottles.' It was just old junk to him."

There it is, progress. Mesquite eradication, historical destruction. Now the mesquite is back and the artifacts lost.

We pulled up to the set of pens where last year I collected my tumble-weed and barbed wire and dried cactus. "Here we are," Gary said as we got out. "I've been here," I told him. Whatever stirrings passed through me last year had been accurate—I'd stumbled upon the site of Old Adobe, headquarters of the JAL. I looked around again, this time with certainty: set of pens, old tin shed, metal storage tank, wreckage of a wooden windmill. Not a tree standing anywhere. Around us arid land rose gently and slightly—we were in the center of a shallow depression—under sharp blue sky. Not a single cloud.

"That enclosure yonder has a big large-bore irrigation pump," Gary told me. "I can imagine that they looked at this and said, 'Aha! This is what we're going to do with this little bowl.' You can understand why the JAL cowboys found water here in this little depression. That's why it is so well situated. It doesn't give you much of a windbreak, but it gives you something."

I could imagine the Cowdens in 1885, when they began developing waterings along Monument Draw, and in 1886 when they built Old Adobe, looking for any shelter available on these treeless plains. A slight dip in the earth would do. So they dug for water and ducked the wind. Here. They hunkered down—cowboys, their wives and children, their horses and cattle. They lived at first out of wagon beds, tents, dugouts. They ate beans, bacon, sourdough biscuits, jackrabbits, deer, antelope, and as little of their own beef as possible—too valuable to eat. A later tale reflected the notorious Cowden thrift: as a jackrabbit raced off through the mesquite and greasewood, a cowhand remarked, "Yonder goes a JAL beef!" Another JAL cowboy claimed you could not work for the outfit unless you could kill a jackrabbit with a rock, your supper.

Thrift and determination were how the Cowdens survived and flourished where so many others failed in this country. For thirty years Old Adobe was the working headquarters of the JALs. From here wagons headed out to work 350 square miles of range. Jean Cowden has a painting of the one-room house, flanked by cottonwoods nursed from land otherwise bare of trees. To one side stands a windmill erected over a hand-dug well. Now all that is gone; not even a photograph remains.

Yet we did find some remaining evidence of the JALs. Spence Jowell at one point took a wagon south from headquarters to collect salt from the shores of shallow, briny lakes that seasonally appear and disappear as summer showers fill them and summer sun bakes them dry. The salt was brought back for JAL cattle, as they watered at various wells and tanks the Cowdens had built up and down Monument Draw. Spence also stopped at the sandhills, where willows grew around small ponds that might recede and disappear but indicated water near the surface. He cut a number of

"MIDLAND COUNTRY"
RANCHES
1 inch = 16.5 miles

Monument Draw

✦ 84 Watering
E. P. Cowden/1905

✦ McKenzie

● Andrews

✦ C Ranch

Old Adobe ✦

✦ Rountree
E. P. Cowden/1928–

✦ Wyche Ranch

Jal ●

✦ Guy Cowden
1928–44

NEW MEXICO
TEXAS

✦ Rube Evans

Llano Estacado

Midland ●

Cowdens/1883

Sand Hills

Odessa ●

Kermit ●

Texas & Pacific Railroad

Monahans ●

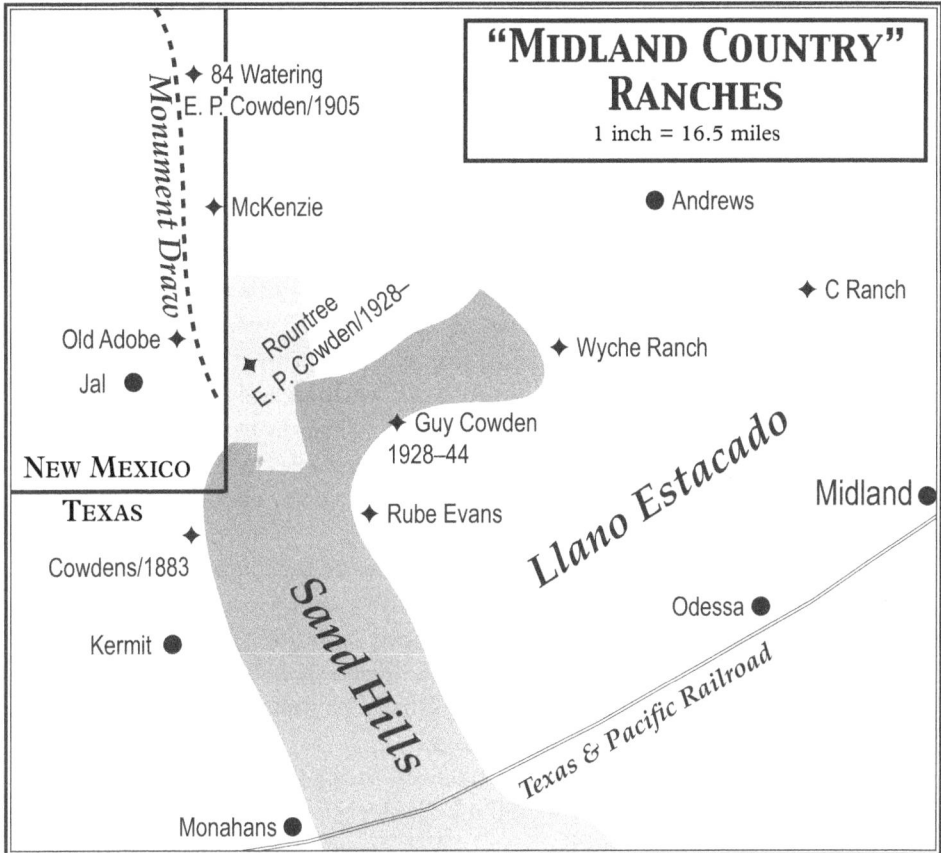

willow saplings for fence posts and brought them back to Old Adobe, where they built a circular set of pens. The posts were set down stockade-fashion, and shortly after they were in place, rains came, and the willow saplings took root again and grew into a living corral. Not even the twentieth century and a bulldozer have erased it completely. There, where they have no business growing, grew willow shoots that Gary and Brian and I could see clearly—the circle is still unbroken.

❋ ❋ ❋

May 3, 2001—Also yesterday, Brian took me by the *Jal Cowboy Sculpture* and gave me the full story. He pitched the project to the chamber of commerce in 1999, got backing, started work: initial drawings, enlargements, templates, and finally, ten-by-forty-foot sheets of quarter-inch steel, over four thousand pounds each. He spent months cutting out the seventeen figures—cattle and cowboys on horseback—at an old building slab nearby. Gene Armstrong then welded drill-pipe frames to the figures, extending six feet below so they could be anchored in holes at the site. Winch trucks and cranes transported and hoisted the cutouts into place on a rise north of Jal. "One cow horn kept hooking the winch truck," Brian said. "We finally got it loose but stuck the truck." Then they poured concrete to secure the figures. The project was finished in March 2000—115 years after the Cowden brothers moved their JAL herd to this area.

The lead rider points south toward the Muleshoe watering, cattle and other cowboys strung out behind him, all of them weathered with rust that complements the landscape. At dusk, with a gaudy sunset behind them—as Brian said, "a frame from one end of the sky to the other"—it's hard to tell they aren't actual life-size silhouettes, so vast is the scale of land and sky. At the site, however, I had to fully extend my arm above me just to touch a cowboy's stirrup.

I suppose that much of the attention first paid to the sculpture has begun to die down, but perhaps—as Brian and everyone else in Jal hopes—the project will have long-term effects. "We need something to get people off the highway and make them want to stop," Brian said. "If people stop and look at the sculpture, then they're more likely to want to stop and eat in town, fill up on gas, and do other things."

Unfortunately, not much traffic passes through Jal. This is, remember, the corner of New Mexico where the U.S. Department of Energy decided to put its Waste Isolation Pilot Plant (WIPP)—a radioactive waste dump thirty-five miles west of here. Although they stick the stuff "2,150 feet underground in a 2,000-foot thick salt formation that has been stable for more than 200 million years," you don't locate such attractions just anyplace—you locate them no place. That's not to disparage Jal; after all, my kinfolks settled there first.

Today I head farther north in New Mexico, to a ranch once owned by my aunt and uncle, Mumzy and Walter Kellogg. Now it's part of the Marsh Ranch but remains connected to the family—Estee Marsh married Luke Kellogg. The land is situated along the Mora River and is wonderful country. With winters too severe for a cow-calf operation like Sam's, the Marsh Ranch is best suited for stocking with yearling steers. It's also a hunting paradise, with trophy deer, antelope, elk, turkey, quail, ducks, and great trout fishing. Oh yes, and a buffalo herd. And *una hacienda linda*— a lovely headquarters compound that can accommodate lots of family.

Not your average cattle ranch, if there is such a thing. Certainly nothing like the sunbaked, windblown JAL Range. I'll be there only for a weekend— my aunt is having a milestone birthday and a surprise party. After that, I change from dress boots to working gear and head for the Cowden Ranch and another branding.

Sin Agua, No Hay Nada

Rain for us made history. It brought to our minds days of plenty, of happiness and security, and in recalling past events, if they fell on rainy years, we never failed to stress that fact.
Fabiola Cabeza de Baca Gilbert, We Fed Them Cactus

Without water, there is nothing. "Sin agua, no hay nada," says Herman, quoting his father, José. Joe Martinez worked for Guy and Rooster Cowden for years, often bringing sons Herman and Albert with him from Santa Rosa out to the ranch. Herman now calls the Cowden Ranch home, as he has for thirty years. He knows the land and labor as well as Sam, and they both know that, in the West, water means everything.

And always has. The JAL Ranch rose and fell on the Cowdens' ability to locate and control water on the open range where their cattle grazed. At first, simple geography protected them—empty, forbidding country that Captain Randolph Marcy reported as "a treeless, desolate waste of uninhabited solitude, which has always been, and must continue, uninhabited forever." The southern reaches of the Great American Desert attracted few competitors and often broke the will of individuals or small outfits that tested the territory.

The Cowdens were nothing if not tough and determined, digging wells by hand along Monument Draw in 1885. May Price Mosley, daughter of an early cowhand, observed the search for water: "Their methods and equipment were of necessity primitive. In digging one of the first wells the hunters drew dirt from the well in a lard bucket tied to a rope which was thrown over the tongue of their wagon for a pulley. Persons never having lived so far from the *Made* things of existence, or from materials with which to make things, cannot realize the handicaps under which these men worked and lived." Methods did not always improve dramatically. Guy Cowden was once lowered into a well on a rope tied to the bumper of a car driven by his wife. Their ranch was in the sandhills, and when it came time to raise her husband, Annie Mae Cowden found the car was stuck in deep sand. Hell of a fix: Guy down a well, car stuck in the middle of nowhere, no help in sight. My grandparents told the story with amusement possible only long after the harrowing fact. "Rock the car," echoed Guy's voice. "Forward, reverse, forward. Rock it." And? "There was no other choice," said Mushy. "I got him out." Ranchers made do and did what they must.

Since cattle graze out from water only so far and return by nightfall, new waterings allowed the JAL Range to expand and endure—sparse ranges soon become overgrazed around limited waterings. New JAL wells were located about ten miles apart, giving each a five-mile radius for cattle to graze out and back, ten miles in a day. (Trail drives, which followed a herd's grazing speed, typically covered six to ten miles a day.) The Cowdens built windmills over their wells, but the first wells in the area were often horse-powered. Plodding a circle, turning the pump, horses or mules were kept awake and active by children, old men, women. Imagine the boredom of man and beast. Robert Rankin remembers, "We had a horse-operated— around and around—pumpjack, pumped that water awful slow. Even before I got four years old, they'd set me down there and I'd keep that horse going round and round. Mother would be at one windmill, and Father would be at another. He'd check on us ever once in a while, and that water pumping a little old bitty stream, and the cattle bawling their heads off, thirsty. It was a very hard life." Hard for everyone: "If there is such a thing as a hereafter," W. C. Cochran wrote, "I doubt if the old timers in this country will ever make the trip, the way they used to punish the old pump horses, working them night and day, and lots of them without a change in twenty-four hours." But sin agua, no hay nada.

Windmills—like barbed wire, another industrial revolution that helped change ranching forever—were introduced to the Midland area at the time the Cowdens first arrived. European windmills were bulky and labor-intensive, not well designed to pump groundwater on the southwestern plains. But after Connecticut inventor Daniel Halladay made modifications, including a vane to direct the wheel into the wind, "American" windmills became fairly simple, self-regulating mechanisms. Southwestern cattlemen saw the benefits of the new technology, as Mosley wrote: "They proved to be the exact key man had been looking for to unlock the fastness of the Llano Estacado." Cattlemen could erect wooden platforms in a day, placing windmills over dug wells and developing widely scattered waterings that didn't require constant attention, just occasional greasing.

And wind, which seems ever-present on the plains, was both resource and affliction for residents. The Great Plains states all offer good to excellent wind energy potential—Texas trails only North Dakota as a potential producer of wind energy; New Mexico ranks twelfth nationally. Different pressure gradients between the plains and the Rocky Mountains to their west help create steady and strong winds. Imagine old maps with gruff clouds blowing over sea or land—above the Rockies one of those characters aims his breath out across the Llano Estacado. Current measurements place average annual wind speeds at thirteen miles an hour for the Cowden Ranch and the former JAL Range—"good" wind energy potential, although anyone who's lived either place knows the wind energy is actual. A cowboy in the open better have his hat screwed down tight or use a "stampede string" knotted under his chin. Otherwise he'll be chasing his Resistol through the sagebrush, cursing the evening breeze.

While wind pumps water from below the surface, it also desiccates the landscape. Along with high altitude, abundant sunlight, low humidity, and high temperatures, wind helps account for dramatic evaporation levels. The evaporation rate given in *The Great Plains* for the JAL Range is fifty-five inches from April to September—in country that receives only twelve and a half inches of rainfall per year. Plants and people better have roots that go down deep, or else they'll dry up and tumble off. The wind giveth, and the wind taketh away.

Between moving sky and arid earth stood the windmill, with the ultimate necessity, of course, groundwater. Despite appearances, the Great Plains from Nebraska to Texas have abundant water. Beneath them lies the Ogallala Aquifer, two billion acre-feet of "fossil" water deposited more than a million years ago. Irrigation for farming in the twentieth century has begun to deplete this vast, invisible reservoir, but for the Cowdens and other early cowmen on the South Plains, the aquifer provided reliable water at shallow depths for watering stock. Cattlemen at first attempted to pump groundwater into long, wooden troughs, but those warped in the sun and wind. The better solution was to dig earthen tanks to hold pumped water; these had to be packed down first, often by herds of cattle trampling the tank bed until it was tight enough to hold water. From such waterings up and down Monument Draw and nearby, JAL cattle grazed out over a spare landscape.

If the problem of drinking water was eventually solved—groundwater pumped by windmills powered by steady plains winds—the problem of

adequate grass was not. Rains to produce grass were not remotely depend-
able, not in country with such minimal rainfall. In the best of times, cattle
needed to be stocked carefully, even when given free range to find grass and
water. Attempting to preserve limited resources, cowboys from various
outfits rode their home ranges to keep their own cattle in place and other
cattle out. It was not always possible.

In the winter of 1884–85 the South Plains were struck by the Big Drift,
when blizzard conditions drove cattle down off the unprotected Llano
Estacado. "George Cowden said there was one hundred thousand head
of cattle passed between Midland and Monahans on the way to the Pecos,"
Cochran remembered, "from as far north as the Canadian River. There
was not anything to stop this drift of cattle until they got to the Pecos.
There was not a wire fence between the Pecos and the North Pole." Cattle
piled up along the river, which had very few crossings and was otherwise
treacherous—its banks steep, current swift, bottom full of quicksand. Not
for nothing had Goodnight called it "graveyard of the cowman's hopes."
The Pecos country, including the JAL Range, became acutely overstocked,
taxing water, grass, and cattlemen. "It took all spring and up until July to
get this drift of cattle back home," wrote Cochran. "There was no place to
water after you left the Pecos until you got to Ward's Wells about seventy-
five miles northwest of Midland and not enough water there to commence
to water the big herds that were driven back from the Pecos. At Ward's Wells
the cattle were all divided and each outfit took their own cattle home."

But not before the Big Drift produced reverberations throughout the
South Plains. Along the Pecos, grazing was destroyed for the near term;
farther north, drift fences were built east-west across the Llano to check
the movement of cattle. That in turn led to the Big Die-up the next winter,
when cattle drifting before new blizzards were stopped by fences and froze
to death in enormous numbers. Some ranches lost up to 75 percent of their
stock. The Cowdens' 1885 move from the Texas sandhills up Monument
Draw into New Mexico was hastened by the Big Drift. Overstocked,
overgrazed country to their south offered no promise, so they developed
waterings in country not so hard hit by drifting cattle and by another,
subsequent disaster—drought.

Droughts in the American Southwest are not uncommon; they've
helped shape the landscape, animal and plant life, and human character
and culture of the region. Those areas of Texas and New Mexico where
Cowdens have raised cattle for 150 years have seen widespread and
extended periods of drought. The mysterious disappearance toward the
end of the thirteenth century of the "Old Ones," the Anasazi culture of

American Indians, while finally inexplicable, is often attributed to extreme drought. Later Pueblo Indian cultures arose along the region's most reliable source of water, the Rio Grande; to its east, the Pecos River supported fewer and smaller settlements. Individual bands of Indians located at scattered springs and creeks so long as water and game were available.

Arrival in the sixteenth century of Spanish explorers did not change the realities and necessities of life in the area. Rather than transform "primitive" Indian cultures that had adapted to the scarce resources of the region, Spanish explorers and settlers more often imitated them. "When the Indians saw our determination to keep to this course [along the Pecos River]," wrote Cabeza de Vaca, "they warned us that we would find nobody, nor prickly pears or anything else to eat. . . . The women carried water, and such was our authority that none dared drink but by our leave." Coronado's expedition confirmed that water, not gold, was the most precious resource in New Mexico and the plains.

When Anglos, the next culture to confront the demands of landscape and climate in the American Southwest, appeared early in the nineteenth century, little was changed since Cabeza de Vaca and Coronado crossed the region. It was still subject to drought, as indicated by written accounts and paleoclimatological records, such as those of the National Drought Mitigation Center. Records from reliable scientific instruments go back only 125 to 150 years, and fewer in frontier areas like West Texas and New Mexico, where climate was critical. So to study ancient climates, paleoclimatologists gather "proxy data" from natural sources; these are the guys who examine tree rings, ice cores, ocean sediments, coral, and other sensitive and revealing natural "instruments." They cut a section from a Douglas fir or piñon pine or juniper and assess patterns in the width and density of annual rings to determine probable levels of precipitation. These they measure against a sophisticated index, the Palmer Drought Severity Index (PDSI).

PDSI records for the JAL Range in Texas and New Mexico go back to 1691 and show individual years, or even decades, of relatively wet or dry conditions. For almost twenty years before the Cowdens arrived, data suggest dry years, if not "severe" or "extreme" drought. Rainfall in West Texas is rarely widespread; a single large cloud might give one part of the range relief and leave another parched. Odds are, the country could not have looked good where the Cowdens began grazing their cattle. And it certainly suffered when drought struck ranges already depleted by the Big Drift. Beginning in the fall of 1885, drought had cumulative effects on land, cattle, and ranchers. "The two years' drouth of 1886 and 1887 broke all the little cowmen on the Pecos," wrote Cochran. "J. V. Stokes says that he stood on the bank of the Pecos in 1886 and counted the dead cattle floating down the river at the rate of forty head an hour. Just imagine what a time we had working that country and drinking that water off those dead cattle in the spring of 1888." In *Crossing Rio Pecos*, Patrick Dearen quotes an 1886 area newspaper report: "The plains west of here are parched and dry, and the carcasses of thousands of cattle are to be seen in every direction. . . . Fully 20,000 carcasses cover the plains. The stench as one passes along the Texas Pacific west . . . is terrible." Here was the South Plains version, the drought version, of the Big Die-up.

Other droughts followed, including extreme ones in 1916–18 and 1934–35, when West Texas joined the Dust Bowl. Imagine Guy Cowden—born in 1900 into a new century but old obstacles, come of age as a rancher in the poor country west of Midland—weathering droughts and wondering if there might not be a better place, with more rain, more water, more grass to raise cattle. His grandfather had come to Texas looking for good ranchland, his father and uncles moved to West Texas and then New Mexico for the same reason, and Guy Cowden too searched for some greener horizon.

When John M. Cowden bought out his brothers' interests in the JALs in 1912, the ranch was in inevitable decline. An influx of homesteaders after the turn of the century had begun to "settle up" the open ranges of New Mexico; some settlers worked for the JALs even as they claimed its territory. They often built houses on section lines, at the four intersecting corners, so that different members of the family could each "live up" claims on land. One section of 640 acres was hardly enough to sustain life in that country; even four sections could not carry many cattle—maybe thirty or forty head, not enough to make a living. Some homesteaders attempted to farm but faced overwhelming odds. West Texas native and newspaperman Barney Hubbs remembered, "Dryland farming, you just couldn't do it. Just wasn't enough rainfall. Maybe one year out of ten, you'd get a half-crop. . . . Some of the homesteaders had windmills; some of them didn't. Some of them couldn't get water, didn't have the money to keep drilling. Homesteaders generally were broke people. They came out here because they were broke, and they could get the four sections of land. . . . The homesteaders ended up, most of them, selling out to the ranchers. . . . The rancher would buy their homestead for a dollar an acre and assume their debt. Lots of the people now that got lots of land got their ranches that way."

May Price Mosley described the period of transition: "Soon after the turn of the century small ranchers—aided by much improved well drills—

began coming in and settling [since it was public domain] in the ranges of the large ranches, faster than the ranching concerns could profitably buy them out. . . . With statehood in 1912, New Mexico selected millions of acres of public domain which were granted to the state in lieu of 'certain concessions' made to the federal government, and on receiving title from the government to the selections made, at once established a lease law on state-owned lands. Thus was some of the old range saved to the ranching industry from the homesteader, but it was no longer free. This protection—since lands so leased could by law be fenced—ended free-grass or open range ranching on the plains which we call the Llano Estacado."

Unless ranchers leased land or bought out homesteads, settlers on public land in New Mexico couldn't be stopped legally. Elsewhere in the West, notably Wyoming during the Johnson County War, big cattlemen attacked "nesters." Such troubles didn't plague the JAL Range, as J. Evetts Haley's notes from an interview with Bob Beverly reveal: "They began surveying se. N.M. after I went to JALs. I went down to Midland, got old man Bill [Cowden] & all others into Bank & told them they were going to have to start spending some money if they were going to hold any of country. Old Man Bill sd.: 'They aint no use spending good money for bad money. We've had the country 32 years. Give it back to the people, boys.'"

If the Cowdens couldn't, or wouldn't, hold together the JAL Range, they didn't abandon ranching; the various brothers and in-laws simply went separate ways. The youngest, Gene, ranched near Kermit, Texas, west of the sandhills, and eventually on a ranch called Rountree, after the headquarters windmill named for Stumpy Rountree, a JAL cowboy who had homesteaded the watering for the Cowdens, always intending to sell it to his employers after "proving up" his claim. Rountree was typical of ranches in the area—some sand country with shinnery, some "tight" land, no surface water, scattered windmills. A 1971 appraisal of the ranch before its sale out of the family noted:

This ranch lies in the southern tip of the High Plains. These are sandy soils lying level to undulating to gently rolling with none of it suitable for cultivation and all subject to wind erosion. Approximately 3,200 acres are classed as Dunesand lying hummocky and billowy and strongly susceptible to severe wind erosion. This type of land is of very little value, producing very little forage. It is practically impossible to get around over other than horseback. The balance of the ranch has a few mixed land sites of a tighter nature but these areas are small and the vast majority ranges from sandy to deep sand sites. . . . Generally speaking this ranch is fairly adequately watered by wells, pipelines and earth tanks. However, it is a long way from the most desirable for good operation. Pastures are large and are far apart, causing cattle to travel long distances to water. Also several wells are gypy and one has been abandoned due to salt water.

"Gypy" water is common to the area, as indicated by the Pacific Railroad surveys conducted in 1854. Geologist William Blake reported: "Captain Pope also mentions immense outcrops of gypsum . . . and states further, that 'numerous caves of pure gypsum of dazzling whiteness within, are found in this gypsum formation, which extends over a distance of 150 miles along the route.'" Drinking water tinged with gypsum—used now commercially for sheetrock—is no treat. Brackish, alkaline "gyp" water in the JAL Range and elsewhere in the Southwest was a bane to travelers and cowmen; getting "gypped" in cattle country referred not to losing money but to the purgative effects of a dose of gyp water. "No one ever went back to get another drink out of a 'gyp' water hole," wrote Carl Benedict in *A Tenderfoot Kid on Gyp Water*, "unless nearly crazed by thirst." Over time, as waterholes or wells suffer repeated evaporation or subsurface depletion, mineral content rises as the basic geology emerges.

Gypy wells on Rountree are but one of the drawbacks to ranching. As the 1971 appraisal report noted, "Other adverse factors directly effecting the overall carrying capacity are not only the virtually worthless Dunesand area of about 3,200 acres, but even worse is the large Dollarhide Oil Field covering approximately 6,900 acres in the northwest corner of the ranch. Much grazing land is lost to well sites, roads, equipment, pipelines." However "adverse" and inconvenient to grazing, the Dollarhide Field and others were the eventual solution to the Cowdens' long struggle as ranchers in West Texas and New Mexico. The prospects of Gene and Tennie Cowden, their daughter Marylee and sons Guy and Al, and subsequent generations were dramatically changed by the discovery and development of the Permian Basin after 1924, as oil followed cattle as the backbone of the West Texas economy, and so helped to preserve ranching.

The most famous oil well in Texas—Spindletop, near the southeastern city of Beaumont—blew in on January 10, 1901. Two days later, the *Galveston Daily News* marveled: "Reports from the well late today say that the stream is spouting fully 175 feet high in a solid form. The overflowing oil has overrun the levees made by the men and is now banking against the Sabine and East Texas right-of-way. . . . Several syndicates purchasing oil lands for development have been formed here today, and several tracts of land have changed hands at what would have been fabulous prices a week ago." A classic gusher, Spindletop was capped, and production began. The field produced 3.59 million barrels its first year, and nothing was ever the same in Beaumont or the rest of Texas. Except that sin agua, no hay nada: "Clean water became difficult to find due to pollution in the area. At one point, water cost five cents a drink, or one dollar a gallon, and oil cost three cents a barrel. Whiskey was promoted as a preferable alternative to water, which created a rowdy and lawless atmosphere."

Oil companies sprang up—Texaco, Humble (Exxon), Magnolia (Mobil), and Gulf (Chevron) all trace their origins to the Texas oil boom. Development of oil fields followed in other areas; among the last, not surprisingly, was the Permian Basin in West Texas—what had been frontier territory was explored only after more accessible areas. The results were dramatic. A geology lesson covering 500 million–600 million years should explain everything.

Before current continents drifted apart, before they first collided, most of western and northwestern Texas was covered by shallow marine waters, with attendant plankton and coral reefs, plant and animal life in turn. To the west stood the Rocky Mountains, to the east rose the Ouachita Mountains, now largely buried. During the Permian period, 245 million to 86 million years ago, many corals and all the little trilobites died, while species like dinosaurs evolved, "soon" to provide lots of organic-rich deposits. Some 100 million years later, the continents pulled apart and the Rocky Mountains ascended further, leading to eventual retreat of the shallow sea covering West Texas. Rivers from the mountains deposited sediment over the landscape as they flowed southeast, as Webb noted in *The Great Plains:* "Once the streams pass beyond the mountain boundaries into the arid land, they dwindle, fail, and deposit their load. Their failure is due to rapid evaporation into the dry air, to absorption into the dry, porous earth, and to lack of local precipitation and augmentation from tributaries." Deep beneath the resultant plain—the Llano Estacado in this case—remain the earlier marine rock foundation, subsequent deposits, and oil.

Discovery and development of Permian Basin oil from 1924 to 1930 were followed by alternating periods of diminished and increased activity, in part because of the area's isolation, in part because World War II disrupted markets: "In the last year of the war, with the completion of the Big Inch and Little Big Inch pipelines, which carried Texas oil to Eastern markets, the industry picked up somewhat. During the course of the war, however, shortages of trained personnel and the diversion of steel for war uses continued to hold back areas of known reserves." Such was the case when "No. 1 E. P. Cowden, drilled to 8,012 feet in 1938, indicated the likely presence of an oil-bearing reservoir" the area. After the war, markets and exploration expanded; in the Permian Basin, major fields at Goldsmith and North Cowden, among others, helped propel Midland into a boom-town. Wells—oil now, not water—were being drilled all over ranchlands, including Cowden land, as Samuel D. Myres noted in *The Permian Basin: Petroleum Empire of the Southwest:* "An outstanding field opened in the Forties is the Dollarhide, named for pioneer rancher Charles Dollarhide, near whose old homestead oil production was opened. Located in extreme southwestern Andrews County, the field extends on westward into Lea County, New Mexico. The discovery of oil in the field occurred in August, 1945."

North of Rountree headquarters, Dollarhide evolved into a "legacy" field, with payloads from five different zones—Clearfork AB, Clearfork C, Devonian, Silurian, and Ellenburger. Myres ticked off staggering numbers: "Production from the Devonian totalled 57,830,277 barrels of oil at the end of 1974. . . . [Silurian] production in the field increased to a total of 31,441,020 barrels at the end of 1974. . . . [Ellenburger] cumulative production through 1974 was 23,674,669 barrels. . . . Appreciable Clear Fork production began in 1949 . . . and the cumulative production at the end of 1974 was 26,007,925 barrels." A barrel of oil measures forty-two U.S. gallons, so the figures become difficult to fathom, especially when you look out over the landscape, the surface mostly barren but for pumpjacks spaced along sandy oil-field roads. Given that all oil fields—based on a declining natural resource—play out in time, Dollarhide remains remark-ably productive, one of the top one hundred producing oil fields in the United States at the end of the twentieth century. The Cowdens were just looking for cattle country.

Most interesting is what the discovery of oil did—and didn't do—for those who had faced the rigors of ranching their whole lives. Discovery of the Dollarhide field came after the Depression and the Dust Bowl, when everyone in the country struggled, even established families like the Cowdens. Guy and Annie Mae married in 1922 and had three children by 1931, Guy working first for his father, then his uncle Charlie Goldsmith, and finally himself. From 1932 to 1935 he leased a ranch east of Rountree, borrowing money—the distinction he made between a cowman and a cowboy—to operate his ranch. A letter dated December 31, 1932, to the bank by inspector John H. Edwards indicates the situation as the national Depression deepened and drought took hold: "The applicant says he did not make his expenses plus interest last year. Says he will need $35.00 or

$40.00 to pay 1932 tax and thinks he can get along on $600.00 for expenses per your offer. He will have nothing to speak of to sell until the fall of 1933 when he plans to sell his calf crop, some fat cows etc. . . . This 35 section ranch is in the southeast corner of the Texas side of the old Jal Ranch. This ranch has been a money maker ever since West Texas was settled. It made the foundation of the great Cowden fortunes accumulated by the older men of that name. . . . Guy Cowden is a young man, bred right, raised right, and is right in every respect. He is a good cow man, lives on the ranch and does all his own work. Admired and respected by all his neighbors. I recommend him as one of the best moral risks you will get."

The inspector's support for rancher and ranch held steady even as conditions deteriorated. In August, 1933, Edwards wrote the bank again: "The elements have not favored him and he will need some additional help. . . . A few showers have fallen on the ranch, which has kept the growth in fair shape in the heavy sand, but with little effect on the eight sections of hard land pasture. . . . Prevailing drouth and no buyers prevented him from doing anything but stay put. . . . The grass is gone from the eight sections and he has an alternate of pushing them over into the sand hills and crowding his calf crop and damaging them or, if possible, to move them off to other range in which event he would need to get an advance of additional money from you." In November 1934, Edwards valued 730 total head of cattle at $21,165 and added: "This is one of your very best loans. . . . Security, place of business and above all the excellent character of the man and woman responsible for the payment, considered."

Excellent character enabled the Cowdens to weather terrible conditions, so that the discovery of oil saw them still ranching and owning land. Gene Cowden ran Rountree with son Al; son Guy had purchased a ranch nearer to Midland. When Dollarhide was developed after 1945, the rigors of ranching, Depression, drought, and world war put sudden money in perspective. However much Guy Cowden and other Midland cowmen might celebrate—drinking hard, buying Cadillacs, and such—gritty ranch life was never far from mind. They all knew that sin agua, no hay nada. Their situation echoed that of Charles Goodnight in 1866, returning from New Mexico after a successful trail drive, yet facing the stern environment along the Pecos and realizing: "Here you are with more gold than you ever had in your life, and it won't buy you a drink of water, and it won't get you food."

Yet no one complained as drilling rigs went up on Dollarhide, lit at night in country where no lights broke vast plains of darkness. Pumpjacks followed, slowly rocking up and down, drawing up Texas crude. With the windfall,

Guy Cowden did what previous generations had done: look for better ranchland. Once again, family helped shape migration. Cousin Elliot Cowden had a ranch up in New Mexico near Santa Rosa, and Guy went to check out the country, some 250 air miles northeast of Midland and about the same distance up the Pecos River from the old JAL Range. The Pecos above Santa Rosa was an entirely different river from the one twisting through Texas toward its junction with the Rio Grande. Lieutenant A. W. Whipple's 1854 *Pacific Railroad Survey* reported: "The Pecos river is here clear and rapid, its waters pure and sweet, forming quite a contrast to those at the several crossings from San Antonio to El Paso, where they are always turbid, brackish, and disagreeable. . . . There, its valley, for hundreds of miles, is a black and dreary waste, with scarcely a shrub to relieve the eye of the traveller; here, its fertile banks are dotted with innumerable small plantations, and towns, so characteristic of New Mexico."

With the Canadian and Pecos and Gallinas rivers watering the country and the Rocky Mountains with their promise of Santa Fe in clear sight, the area offered more than the "dreary waste" of the JAL Range. Yet, like the JAL, the new country had seen passages of many cultures over the years. From the west had ventured Spanish conquistadores hunting for cities of gold and later Hispanic pastores and ciboleros and comancheros visiting the plains. From the east, along the natural concourse of the Canadian River, Apache or Comanche raiders rode to attack Indian pueblos and Spanish settlements of the Rio Grande. Later came American parties: Josiah Gregg in 1839–40, Captain Marcy in 1849, and Whipple's survey in 1853.

Of U.S. government crossings, Marcy's was the first with details of likely travel across the Cowden Ranch. Accompanying Marcy—in addition to the wagons of gold rush emigrants, for whom the military escort needed to provide reasonable passage, pasture, and water—was Lieutenant J. H. Simpson of the Corps of Topographical Engineers, looking for a roadway to California from the states. Along the Canadian River, Simpson noted "unmistakable signs everywhere existing in the shape of numerous and well beaten trails, if not of cart roads, concentrating at this point, to show that the place where we crossed it has, from time immemorial, been a grand encamping ground for Indians and Mexican traders." In passing, Simpson felt "justified in engrafting upon [creeks] such names as to me, from the circumstances connected with them, seemed most appropriate," including "Hurrah Creek . . . because, it being the first affluent of the Pecos we came to, it gave us certain indications we were approaching the close of our journey." Hurrah was Esteros, and their camp most likely near the present Bar Y headquarters. Marcy's initial diary entry referred to the creek as "Gallenos," although his full account makes the locations and waters clear:

> *June 22.*—We travelled over a smooth dry prairie to-day for thirteen miles, and encamped upon "Gallenos [Esteros] creek," an affluent of the Rio Pecos. Here we had good water, with some cottonwood upon the creek, but the grass is short, the Mexicans having recently made this a pasture ground for large flocks of sheep. We have passed many high sandstone hills on each side of our road to-day, and we have seen far off in the distance the northwest mountains, with their tops covered with snow. The country in this vicinity is a miserably sandy plain, and fit for no other purpose but for grazing sheep.
>
> *June 23.*—Our road passed for fifteen miles over a very high rolling prairie, with detached rocky hills upon our right, running off towards the Canadian, until we reached the principal branch of Gallenos creek; a fine running stream, with a rock bed, and fifty yards wide. Nine miles from here there is a spring of cold water [Chupinas]; and at this place the road forks, the right leading to San Miguel, the left to Anton Chico. We took the latter, and reached the Pecos before night, making a drive of thirty one miles.

A route from the Bar Y northwest across the Cowden Ranch along Esteros Creek, through the natural saddle in the mesa to Alamito, then west to Las Chupinas spring and southwest to Anton Chico, exactly conforms to the descriptions and distances of Marcy's account. His Delaware Indian guide, Black Beaver, had worked for the American Fur Company for ten years and knew the route, undoubtedly the one commonly used by trappers and traders and Indians.

As Marcy noted, this was also the route of pastores from Las Vegas and other villages, who brought flocks of sheep to graze en route and upon the Llano Estacado beyond. Simpson added, "I was in the midst of the largest flock of sheep I had ever beheld. . . . According to the shepherd's statement, 2,000 head, and including another flock near, said to contain 4,000, the whole aggregated some 6,000." This was also the path that ciboleros traveled from Hispanic New Mexico to the plains to hunt buffalo, as Josiah Gregg observed: "Every year large parties of New Mexicans, some provided with mules and asses, others with *carretas* or trucklecarts and oxen, drive out into these prairies to procure a supply of buffalo beef for their families. They hunt like the wild Indians, chiefly on horseback and with bow and arrow, or lance, with which they soon load their carts and mules." Gregg encountered other Hispanic adventurers: "Parties of *Comancheros* are usually composed of the indigent and rude classes of the frontier villages, who collect together several times a year and launch upon the plains with a few trinkets and trumperies of all kinds, and perhaps a bag of bread and maybe another of *pinole,* which they barter away to the savages for horses and mules." Livestock that comancheros received from Comanche raiders originated from many sources—other tribes, Mexican villages, frontier ranches like that of the Cowdens back in Palo Pinto. The route that Marcy took through spare and sparsely settled country was a major thoroughfare for three different cultures.

Whipple's exploration followed that of Marcy, whose report was an important resource for government and private parties both. In 1853–54, under the direction of Secretary of War Jefferson Davis, surveys to "ascertain the most practicable and economical route for a railroad from the Mississippi river to the Pacific ocean" were ordered for a number of latitudes, including the 32nd and 35th parallels—through what became the JAL Range and the Cowden Ranch. For political reasons, Jefferson Davis favored the most southern route, which also offered advantages of topography and climate, but the Civil War would intervene before any line was selected. Whipple's survey along the 35th parallel was nevertheless extensive and informative, as when he described his camp on Esteros Creek, at the Bar Y or perhaps slightly farther downstream:

> *September 23—Camp 51.*—Set out at 7 1/2 A.M.; the morning calm and beautiful. The road ran along the southern base of a red sandstone bluff, worn into curious shapes resembling monuments, vases, and caves. . . . We crossed the spur of a hill, which, by a detour to the right, might have been avoided, and entered a prairie interspersed with dwarfed cedars, cacti, yuccas, and mesquites. From the spur was a fine view westward, over mesa hills, to a blue sierra, said to belong to the Rocky mountains. . . .
>
> *September 24—Camp 52.*— . . . The morning was cool; the thermometer, at sunrise, standing at 43°. By an unusually steep ascent, we climbed a long, low ridge of hills to the crest of the summit dividing the waters of the Canadian from those of the Pecos. We were then nearly on a level with the top of the Llano, which appeared a mile or two to the left. . . . Passing a pond of water bordered with green grass, we proceeded to l'Assisteros [Esteros], thirteen miles, and encamped. This is the first affluent to the Pecos, and is styled by Simpson [during Marcy's 1849 journey] Hurrah

creek. It flows through a fine valley of buffalo-grass, that appears to be now in its most perfect state. It is quite a rapid stream, eight feet wide, with cool, pure water. We camped at one o'clock, though there was no wood in sight, for the mules needed rest.

Near camp is a hill of sandstone, with masses broken into singular forms strewn upon the sides and at the foot. Among them are enclosures, and slight walls have perfected the seeming intention of nature—rendering the place quite a fortress. I believe that the New Mexican shepherds secure themselves and their sheep here from Indians and from wolves. . . .

September 25—Camp 52.—Our cattle becoming foot-sore and weary, to avoid leaving many of them behind, we determined not to move camp to-day, but allow them to rest. Dr. Bigelow immediately set out upon a botanical excursion. About four miles southwest he struck the bend of a river, probably the Gallinas [likely the Pecos at Horseshoe Bend], flowing in a cañon cut almost perpendicularly through the plain to a depth of about 500 feet. . . .

September 26—Anton Chico.—Took an early start; watching, as we went along, the varying tints of approaching day, so beautiful in this climate. A mile from camp we crossed the sandstone bed of the second branch of Hurrah creek, where there was water in pools. Commencing the ascent of a prairie ridge, we rose gradually, in three miles, about 400 feet, when we found ourselves upon the limestone surface of an extensive plain. For a mile or more the road was smooth, and then succeeded rolling prairie, beyond which was a forest of dwarf cedars. Fifteen miles of survey brought us to the Rio de Gallinas, a creek with pure running water, but with neither wood nor grass upon its banks. Passing over an undulating country, we reach the crest of a hill overlooking the Pecos at Anton Chico.

Additional details from the final report give more specifics, as Whipple's party seemed to travel over the heart of the Cowden Ranch: "From Camp 52, on Hurrah creek, to Camp 53, at Las Chupinas, is a high broken prairie, intersected by many ravines. . . . Rio Gallinas, . . . whose outlet is south into the Pecos by narrow defiles near our line, is a beautiful, bold, clear, running stream, affording water at all seasons of the year, while the Tucumcari and Pajarito creeks, in the immediate vicinity of the Tucumcari hills, flowing north into the Canadian, will afford water doubtless nearly all the year." In locating a railroad line, as in every exploration and enterprise, water was critical.

While the transcontinental railroad in 1869 followed a route well north, other railroad lines were built near the 32nd parallel, but not the 35th until after the turn of the century. Congress in 1866 chartered the Atlantic and Pacific Railroad Company, whose route from Missouri to San Francisco proposed to follow the 35th parallel, but the company never completed its track. A map of the right-of-way shows the proposed railroad running along the common border of the Antonio Ortiz and Anton Chico land grants, just inside the north line of the current Cowden Ranch. Along with topography, water available where Little Conchas and Alamito windmills now operate helped define the route.

Beckoned by the reports from his cousin, Guy Cowden in 1948 went to Santa Rosa, situated on the Pecos just beyond the northwest rim of the Llano Estacado. Farther north, a portion of the Bar Y Ranch was for sale, in the area touted by stock commissioner J. H. Koogler in *Illustrated New Mexico,* an 1885 publication of the Bureau of Immigration:

> Between [the Pecos and Canadian] valleys are vast areas of high, undulating country covered with the native gramma grass, rich and nutritious, which forms the pasture land for stock. The eastern and southern portions of [San Miguel] county, watered by the Canadian and Pecos rivers and tributaries, are wonderfully fine stock growing regions, combining water, grass and shelter. The best stock ranches of the Territory are situated there, and the herds are steadily increasing in number and value. . . . This is preeminently a breeding country, owing to the mild and equable climate, and cattle multiply with a rapidity not experienced in more northern latitudes.

The Bar Y, owned by the Driggers family, had a complicated history of ownership. After Indian tribes were dispossessed of their traditional lands, and following Mexican independence, settlers in 1822 successfully petitioned for a land grant to form the village of Anton Chico, including 370,000 acres for grazing stock. The village of Anton Chico was abandoned because of attacks by Comanches but was resettled in 1834; the grant itself became part of the contentious relationship between Hispanic and Anglo interests following the Mexican War, as American cattlemen sought to displace Mexican sheepherders on the grant's common lands. Lawsuits continued into the twentieth century, and not until after New Mexico became a state was the case finally decided, when Anton Chico claimants settled out of court and the New Mexico Land and Livestock Company

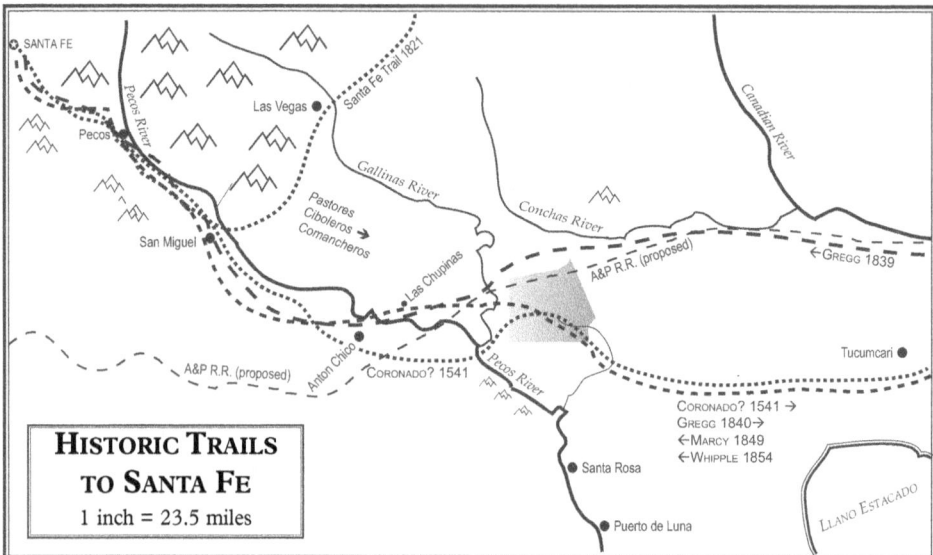

HISTORIC TRAILS
TO SANTA FE
1 inch = 23.5 miles

SANTA FE

Pecos River

Las Vegas

Santa Fe Trail 1821

Canadian River

Pecos

Gallinas River

Conchas River

San Miguel

Pastores
Ciboleros →
Comancheros

Las Chupinas

A&P R.R. (proposed)

← GREGG 1839

A&P R.R. (proposed)

Anton Chico

CORONADO? 1541

Pecos River

Tucumcari

CORONADO? 1541 →
GREGG 1840 →
← MARCY 1849
← WHIPPLE 1854

Santa Rosa

LLANO ESTACADO

Puerto de Luna

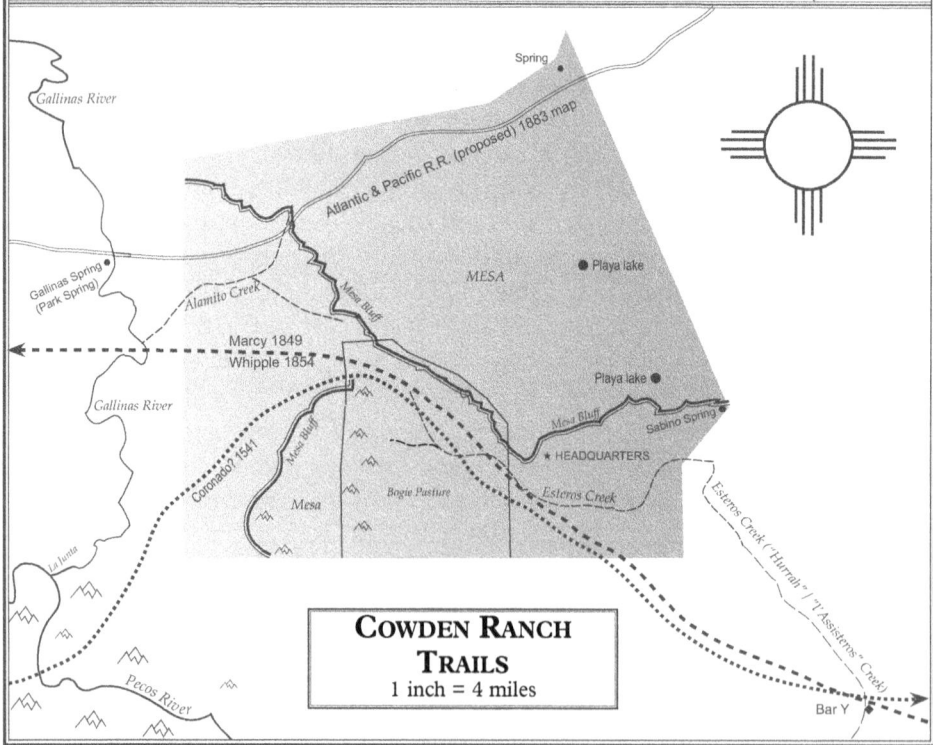

COWDEN RANCH
TRAILS
1 inch = 4 miles

Gallinas River

Spring

Atlantic & Pacific R.R. (proposed) 1883 map

Gallinas Spring
(Park Spring)

Alamito Creek

Mesa Bluff

MESA

Playa lake

Marcy 1849
Whipple 1854

Gallinas River

Coronado? 1541

Mesa Bluff

Playa lake

Mesa Bluff

Sabino Spring

★ HEADQUARTERS

Esteros Creek

Esteros Creek ("Hurrah"/"7 Assistens" Creek)

La Juita

Mesa

Bogie Pasture

Pecos River

Bar Y

took possession of most grazing land. The Preston Beck Jr. Land Grant, an overlapping claim granted in 1823 by Mexican authorities, further complicated titles. The abstract of title filed in November 1948 for the sale to Guy and Annie Mae Cowden runs almost four hundred pages and includes documents in Spanish, mining locations, warranty deeds, lawsuits, leases, and quitclaim deeds.

Whatever its troublesome legal history, the country was in great shape when Guy Cowden purchased it—high plains green with grama grass, water running in the creeks and holding in numerous shallow lake beds called playas, from the Spanish for "beaches" or "shores." Playa lakes are shallow, flat basins common on the Llano Estacado; from 17,000 to 37,000 are estimated to dot the landscape, sometimes holding water, other times dry. "Time of the lakes" was plains lingo indicating that water was available in otherwise dry playas; cattlemen could then drive herds across the Llano. Playas remain critical in recharging the underlying Ogallala Aquifer; they may go dry but nevertheless contribute to the viability of life on the plains, as water percolates down without runoff and erosion. At the time Guy Cowden considered buying the ranch, the playa called Big Conchas was full. "I mean plumb full," Sam says. "They talked about buying a boat!" The abundant water was no illusion, but neither was it permanent.

Cowden Ranch headquarters lie at approximately latitude 35° north, longitude 104° west, the coordinates of one cell of the Palmer Drought Severity Index. According to the PDSI, that cell had suffered only one year of drought in all the 1940s, which was balanced by a number of years of excellent rainfall and growth. How could Guy Cowden fail to appreciate the difference between this well-watered country and the dry, brushy rangeland around Midland? And how could he know what was coming?

In the decade that followed, the same ranch that had looked so promising entered a period of prolonged drought, "moderate" to "extreme" every year from 1950 to 1957, with "critical" conditions in 1954. Annual rainfall totals from Santa Rosa, although subject to widely variable local precipitation, support the PDSI. Every year from 1950 through 1955 had well below the median level of 14.56 inches per year: 10.62 inches in 1950, 8.70 in 1951, 10.83 in 1852, 7.46 in 1953, 10.71 in 1954. And in 1955, after five already bad years, Santa Rosa received its lowest recorded annual rainfall, only 6.63 inches. Less than seven inches of rain is liable to ruin any rancher. The sequence goes: no water, no grass, no cattle.

Statistics give information but don't tell the whole story. The story goes that my first word was *agua*. That may be fact, may be legend, may be myth.

But history, especially family history, is often mythic. Was I thirsty? Simply learning to speak the world around me? "Agua," I said, an infant pointing at a glass of water faintly red, colored by the earth through which it rose. *Colorada* it would be in Spanish, *agua colorada*—red water. When we visited my grandparents in New Mexico, I had a Spanish nurse, a *niñera* from the village of Puerto de Luna, south of Santa Rosa. I seem to remember her house, a small adobe dwelling on a dark arroyo, lamplit, the night black and wide beyond the lambent window glass. Memory is an act of the imagination, though I know she cared for me, speaking Spanish, its syllables running like water in a dry land. Agua.

One irony of our summer trips to the ranch, where water was so scarce and precious, was how often we suffered from it, how many times we got stuck on the ranch road, in many different locations. A brief, hard shower might turn a low stretch of red ranch road into a slick and/or impassable scramble, a wild ride through the Wild West. And if you got stuck bad, you were stuck nowhere. Trying to ease your vehicle out of a greasy clay mire was difficult—thick, heavy rims of red mud forming on your boots, sticks and stones intended for traction simply disappearing under spinning wheels—but failure might mean a ten-mile hike to the ranch house.

More often, though, the drive was dusty, through shortgrass country rarely green, except around water tanks or playa lake beds. There were no real rivers to cross, although the Pecos cut breaks in the country just to the west. A water project established a dam on the Pecos, creating Santa Rosa Lake for conservation and flood control. The thirty-eight-hundred-acre lake in the middle of nowhere would glitter in the distance like a mirage, an oasis you might blink away in an instant. Beyond it, the ranch road crossed just one creek, twice, its bed often dry, the last crossing on the Cowden Ranch itself. There's a family legend about that too.

It was 1950, just after my grandfather bought the ranch, and before improvements of any sort. We were there in summer for a visit, possibly our first look at the new place. I can't say, since my mother was pregnant and I was still floating in amniotic fluids. I'm told we were all there—father, mother, sister, unborn brother—at headquarters north of Esteros Creek, which "runs" through the lower ranch. The Spanish word *estero* means, among other things, "swamp" or "stream," although Esteros was more often a gully, an arroyo with occasional pools of muddy water lingering from the last rain. The story goes that a storm came up, one of those enormous summer thunderheads that fill the western horizon, rising like a bad bruise high into an otherwise blue sky. Which worried my mother, twenty-six at the time, in good health but very pregnant. No matter that she had grown up on ranches and had in her character great reservoirs of determination and spirit. She wanted to have her baby in a hospital, with a doctor, not on a ranch, not thirty miles from the nearest small town. She wanted to go back to civilization, to Texas.

"God willing and the creek don't rise," they say, which was the danger— a thunderstorm with what Pueblo Indians call "male rain," a hard down-pour. Even a gentle "female rain," perhaps two-tenths of an inch but still producing runoff, could create an impassable creek. Rain male or female, my mother did not want to get stuck. So my father and grandfather went down the road from headquarters to the creek bed and began tossing in rocks. Flat rocks, round rocks, rocks of any and all sorts—Permian rocks, Triassic rocks, Jurassic rocks. My father was a petroleum geologist, and I can imagine him, even as he worked to make the crossing certain for his wife and unborn child, checking out the geology of his materials. Family legend has them working tirelessly, like volunteers along swollen rivers back east, filling and stacking sandbags. History has them getting my mother out, off the ranch, back to Route 66, and back to Texas, where I was born— a child of August, a child of the sun-blazed southwestern plains—but more than anything, agua marking my birth.

Following my father's work in the oil business, we moved from Texas to Louisiana, to New Orleans, where it rained all the time. All the time. Bayous and rivers were everywhere, the Mississippi rolling down toward the Gulf of Mexico, massive cloud formations rolling back up from the same waters, a constant circulation of moisture I took for granted. Except for summers when we went back to West Texas and New Mexico. My grandparents would invariably ask if we had brought rain—average rainfall in New Orleans is approximately sixty-two inches a year, four to five times what the Cowden Ranch receives. In New Mexico the wettest months are, surprisingly, July (2.39 inches) and August (2.71), when the region experiences a monsoon season. Often associated with the Indian Ocean and Southeast Asia, monsoons are more generally any seasonal winds blowing onshore from the southwest and producing heavy rains. As summer begins in North America, moisture from the Pacific begins to flow northeast over the deserts of Mexico toward the southern Rocky Mountains. When this moist, hot air encounters the cooler barrier of the Sangre de Cristo and other ranges, frequent showers result. Mountain areas receive the most rainfall; locations on the eastern slopes of the Rockies and nearby on the Great Plains—rain-shadow regions—simply hope for the best. If you miss your rains in July or August, you can face serious conse-quences, since the high elevation, constant winds, and minimal vegetation contribute to high rates of evaporation. On the eastern slopes and the Llano Estacado every shower, every cloud in the sky, is valuable.

I remember my grandfather telling the story of some old ranchers sitting around in town, in the shade of a porch, their straw hats off in the summer heat. As they sat bareheaded, looking to the skies for the least sign of a cloud, pigeons flew in and out of the porch rafters. One of the boys—"boys" even though they might be sixty or seventy years old—suddenly sat up, a splatter of pigeon shit dripping down his forehead. As all his friends broke apart into laughter, he just wiped his head and said, "Oh well, anything for a little moisture!"

In the same reverent spirit is a photograph my grandfather had made of some Cowden Ranch cattle. Around a small water hole on the bare, shortgrass plains with a low mesa in the distance, white-faced Herefords stand under a blazing blue sky into which are painted a few white clouds. Like desert mirages, clearly unreal, bogus clouds hang in the air as evidence of a lifetime's longing. "Oh buttermilk skies, oh buttermilk skies," I recall my grandfather singing, "I wish you'd turn to watery milk."

So whenever we arrived at the ranch after our long drive from New Orleans—our car coming over the horizon, producing a rising plume of dust—we came as emissaries of rain, agents of the cumulus and cumulo-nimbus. Indians of New Mexico and the desert Southwest, less scientific, have their own totems of rain—frog or snake or turtle, dragonfly or toad—and hold them holy. But a grandchild from liquid Louisiana might just as well bring rain. We thought of ourselves as lightning rods, sources of a shower, a cloudburst, a soaking rain. We came hoping to please, yearning for rain, learning to watch after supper for some sign on the wide horizon that relief was on the way.

One summer at the ranch I brought no rain. A New Mexico Cattlemen's Association tour was scheduled to visit, a demonstration tour. I remember the events but not the year—was it 1956? I was old enough to be assigned the job of opening Cokes and Dr Peppers for the visitors. A long caravan of pickups and cars was headed first to the Bar Y, then to the Cowden Ranch, where cattle were on display in corrals at headquarters and in pastures along the road. Trouble was, it had not rained in months. The grass was withered, the cattle were poor, and the ranch road had turned a fine deep powder that rose as billows of ruddy dust with every passing vehicle. Dust coated cholla cactus and prickly pear beside the road, leaving the whole country a terra-cotta color, like the coat of a sorrel horse. Each breath you took was gritty, and sand worked into your clothes and skin, abrasive and unrelenting. And that was before the cattlemen came, like a stampede over the southern horizon.

My grandmother could only imagine what fifty vehicles might do. It would be a dust storm; it would be disaster. Dust storms, "black blizzards," had stuck before. They required dry times and high winds, both prevalent at the ranch. I remember the first I ever experienced, alerted one windy afternoon by my grandmother's subdued but horrified "Oh no! Oh no!" After which she scrambled from window to window of the house, closing each one tight, pulling shut the curtains, trying to seal off what was coming. To the southwest I could see an ominous dark wall of dust, miles long and hundreds of feet high, tumbling across the country. The only thing I'd ever seen like it was, ironically, big gray-green waves breaking along beaches of the Gulf of Mexico.

We needn't have worried, however, about disappointing the cattlemen's association; every man and woman on the tour had suffered through similar weather. None had a discouraging word. The country forges durable goods or casts them out.

Fortunately for the Cowdens, the long drought of the 1950s relented, and the country and climate generally reverted to the promise Guy Cowden had first observed. The weather moderated, and though always in danger of localized drought, the country has the ability to thrive with little moisture. The Cowden Ranch in particular is well suited to withstand dry weather, as archaeological evidence shows. Within the ranch boundaries, Indians—Pueblo, Apache, and Comanche—first took up residence. Jean Cowden has assembled an extensive collection of arrowheads found at various sites around the ranch. She's expert in spotting them, lying among countless unshaped stones, for she knows where to look—wherever there is, or was, water. The edge of the mesa is a source of springs and seeps, surfacing from water that has percolated down through the tableland above. Over the years, on trips to these springs, Jean found most of her arrowheads, proof that Indians were no strangers to the area, with the good sense to locate in protected sites near water. The water that first attracted those Indians now reveals their history, since arrowheads tend to surface after hard rains, washed from spots where they've rested for centuries.

Spanish and other settlers later built rock houses, now in ruins, near the same places where Indians watered, showing the same common sense. Sheepherders were probably the first to build, using abundant native stone for their small cabins and corrals, placing them near reliable springs, like Sabino Springs on the extreme eastern point of the ranch. Tucked back into a seam of the mesa is a spring pool fifteen feet across, surrounded by water plants, shaded by the mesa walls from fierce summer sun and drying winds. The spot is always cool and feels like a sanctuary—surrounding piñons and cedars lift twisted branches like a choir, frogs and grasshoppers joining in. After crossing miles of open plains, savvy Indian or Spanish or Anglo travelers would have seen this mesa edge as refuge, noting the trees, knowing they meant water. Searching along the rimrock, they would have found the spring and joined the congregation.

Elsewhere on the ranch are other sources of water. Back west along the mesa, another spring supplies water for headquarters—Sam and Kathy's house, the Martinez home, the barns and corrals. Guy Tom put in a water line in 1988, after windmills at headquarters finally played out. The House Spring doesn't provide an overabundance of water—only one-quarter to one gallon per minute, depending on the weather—but you take what you can get. The water is stored in tanks at the house and used sparingly, as needed. I remember the odd pleasure I felt as a child on visits, because we could bathe only once a week. My two sisters and I all had to use the same bathwater. I was always last in the sequence, but even before anyone washed off the dust, the water ran red. After the tub drained, red sediment lay on the white porcelain, a little creek bed visible where the last water had flowed away.

If we didn't want to wait for that weekly bath, we could always take a swim in one of the cattle tanks. Anywhere from ten to twenty feet across and three to five feet high, the open metal tanks stood beneath the spires of windmills located around the ranch. Attached were smaller tanks where the cattle watered, the ground around them packed hard or occasionally, after rains, a mucky gumbo. Inside the tanks was water, all right, but the bottom and sides were slick with long strands of green algae—nothing like swimming pools back in Louisiana. Tadpoles and minnows come from God-knows-where made the experience all the more unsettling. And remarkable. To a child from New Orleans, what could surpass a solitary dip in a cattle tank on the open New Mexico range, with the windmill above turning and creaking, cold water surging and pausing, surging and pausing, as it flowed from the pipe? I recall feeling absolutely singular. At every windmill was always a tin cup hanging on a nail, available for anyone passing by with a thirst.

Now Sam Cowden operates just seven windmills. Coyote Mill at the southwest corner of the ranch is four hundred feet deep and "weak."

Another weak mill on the west side—where I swam as a child—sits in the Alamito horse pasture. "Fifty feet deep, and weak," Sam says, "but with good soft water. Pumps about one-half gallon per minute." On the south side of the ranch, below the mesa in the Esteros Creek drainage, are Poso, Little Poso, and Creek pasture mills. Poso and Little Poso are each sixty feet deep; Creek pasture mill, farther downstream, is eighty feet deep and strong. All three mills benefit from both the creek bottom and the hydrological movement from the mesa above.

Two other windmills farther on the south side were abandoned. Caps mill, says Sam, was "too deep and too weak and a pain in the ass." So was Bogie, an astounding eight hundred feet deep and "too hard to work on," according to Sam. "It was easier to pump water than work on a mill that deep. Because the ranch is so far from supplies and services, Sam, like his father and grandfather before him, needs to be as self-sufficient as possible. A windmill crew gets expensive real fast, especially if it must travel a hundred miles or more before beginning work on a mill. Bogie mill no longer pumps water.

Up on the mesa is another abandoned mill, Big Red, shut down for the same reasons. Still operating is Antbed, two windmills one hundred feet deep that sit on a fence line and serve three large pastures on the north side. Sam has run water lines from Antbed to "drinkers"—small tanks like water fountains for cattle—in various pastures. He's able to run lines because the windmills are supplemented by an electric pump powered by a gasoline generator, so it can pump 24/7. Beyond Antbed, on the far northern tip of the ranch, is Little Conchas, a beautiful little place near the Conchas River breaks. An old wooden mill surrounded by ancient cottonwood trees, Little Conchas was dug by hand and sits over a good, sweet spring.

In addition to windmills are wells where there's both water and electricity to power pumps. Alamito horse pasture has a "strong, four-hundred-foot well" with a 1.5-horsepower electric pump and provides ten gallons per minute of hard water. It serves a number of pastures on the west side and waters the house and barns at Alamito. A well in the Shipping pasture provides water to the southern pastures, the shipping pens, and stock at headquarters. Wells with electric pumps can be less problematic than windmills, but Sam notes, "You still have miles of pipeline, drinkers to repair and clean out, storage tanks, electricity to pump the water." There's no easy solution, and no alternative, to water.

The best answer, of course, is rain. A good rain in spring will kick off the grass, allowing it to extend its root system and withstand dry periods. Rain also fills the dirt tanks that the Cowdens have dug at various places to utilize runoff. Depending on the drainage and soil composition, some hold more water and hold it better than others. Many of them are what Sam calls wet tanks, good only when there's been a recent rain. Other tanks, deeper or with better drainage supporting them, hold water in all but droughty spells. Up on the mesa are the two shallow playa lakebeds, and Sam has dug tanks into the center of these so they hold some water much of the time.

In the summer of 1998, I returned to the Cowden Ranch for the first time in many years. It was a fortunate decision. The area had received adequate rainfall, as it had when my grandfather first saw the country. In 1997, Santa Rosa rainfall totaled almost thirty inches, twice the average. The grass, while always short, was abundant and green. Water shined in tanks and creeks; the cattle were fat. I saw much the same in the summer of 1999, after the ranch had received good spring rains—an uncommon occurrence, possibly attributable to La Niña. Four inches of rain fell in April, two and a half more in May. Never had I seen it so green. It made a believer out of me, although I had never lost faith in the ranch. Nor had Sam, despite difficult stretches. However hard the cattle business might be, the land continues to sustain Cowden cattle, so long as it is respected and well managed. One or two good years of moisture might mislead some ranchers to overstock their ranges, but Sam realizes how fragile the balance is between climate and country and cattle. And like his ancestors, like every Cowden from the JAL Ranch forward, he knows to pray every day for rain. Sin agua, no hay nada.

Cowden family, Palo Pinto County, Texas, circa 1878. *Left to right, standing, rear:* George Edgar (?), Cynthia Catherine, Annie Lee, John Motherwell (?), Charles Webster. *Middle:* William Hamby, Liddon, Caroline Liddon, Willie Jane, William Henry (?). *Front:* Eugene Pelham, Rorie Wynne. *Not pictured:* Mary Josephine and Nannie Liddon, who died in 1868 and 1867, respectively. Courtesy of the estate of Paralee Payton McMahon.

Front gate, Cowden Ranch, Guadalupe County, New Mexico, May 2000. Looking north across Shipping pasture toward mesa. Cowden headquarters are located at left, just below mesa. Photo by author.

Cowden family, 2000. *Left to right:* Sam, Guy, Abby, Hannah, Kathy. Photographed at Jean Cowden home, Santa Rosa, New Mexico, by author.

JAL Cowboy Sculpture, by Brian Norwood, Jal, New Mexico, May 2002, with author posed in front. The sculpture cowboy is pointing toward Muleshoe Watering on the old JAL Range, which later became the town of Jal. Photo by Brian Norwood.

Old Adobe in Monument Draw, New Mexico. Painting by Flora Farnsworth; in possession of Jean Cowden. Copies are located in the Woolworth Library, Jal, New Mexico.

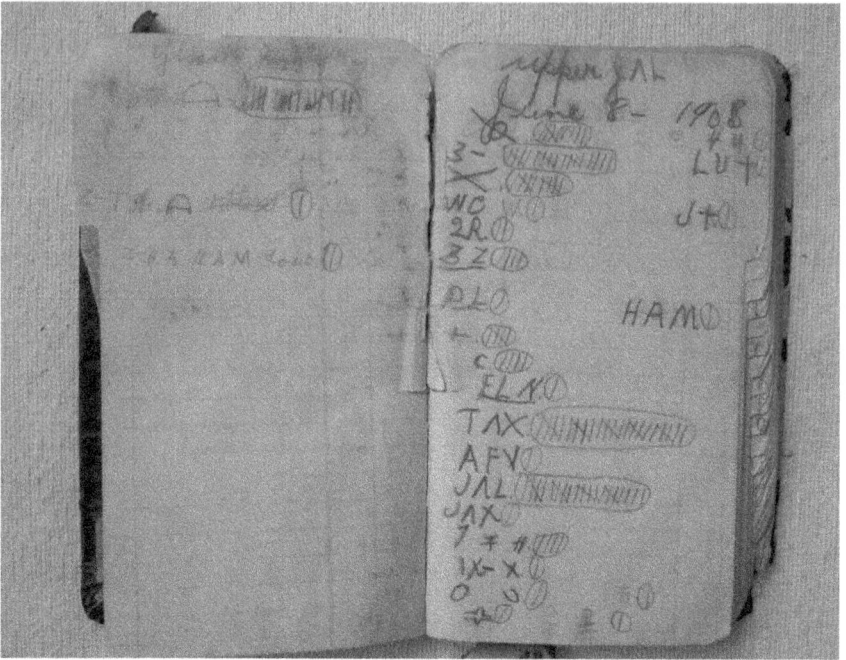

JAL tally book. Eugene Cowden became tally man for the JAL Ranch. The page reflects cattle gathered at Upper JAL watering, New Mexico. The TAX brand originated with Frank Divers and his Dug Springs Ranch, north of the JALs. Tally book courtesy of Jean Cowden.

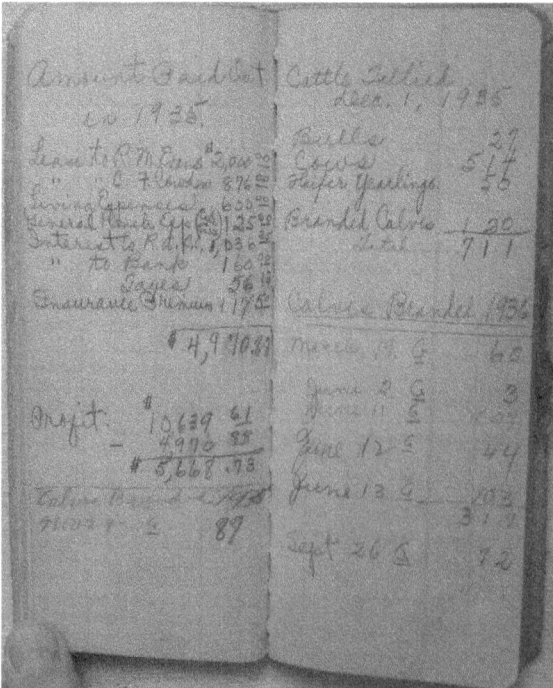

Guy Cowden's tally book, recorded by Annie Mae Cowden. R. M. (Rube) Evans and C. F. (Chunky) Cowden were relatives. Guy and Annie Mae lived south of Rountree in Winkler and Andrews counties, Texas, from 1928 to 1944. Their older children, Gene Ann and Guy Tom (Rooster), were taught at home or on neighboring ranches until the youngest child, Mary Lee (Mumzy), was old enough to start school in Midland. Tally book courtesy of Jean Cowden.

Dollarhide oil field, Andrews County, Texas, May 2000. Named for a former homesteader, Dollarhide began production in 1945 and eventually encompassed seven thousand acres of oil and gas wells. The E. P. Cowden Ranch was referred to as both Dollarhide and Rountree, the ranch headquarters named for former JAL cowhand Stumpy Rountree. Photo by author.

Creek windmill, Cowden Ranch, New Mexico, 2004. In background on right are ruins of a rock housed used by nineteenth-century sheepherders; Esteros Creek runs between windmill and ruins. Springs at mesa edge *(left background)* and runoff feed Esteros and supply ground-water for windmills. Better water and grassland led Guy and Annie Mae Cowden to move ranching operations from West Texas to New Mexico; oil revenues enabled them and others to improve ranches. Photo by author.

Sam Cowden on Dunnie, heeling Buford the mechanical steer, May 2000. Buford allows Sam to practice without a roping partner or calves. Note how his roping saddle pushes Sam forward into good roping position. Photo by author.

Sam Cowden and Hollywood, with Hannah, Abby, and Guy Cowden, Santa Rosa, 1992. Santa Rosa Roping Club championship saddle is on Hollywood, "best horse I ever owned." Photo by Karla White.

The author and Herman Martinez at the Cowden Ranch "Bullpen," May 2000, after branding at Alamito pens. Photo by Kathy Cowden.

Guy Coley Cowden and Seth Shanklin, the next generation of ranchers, at Cowden Ranch, May 2000. Photo by author.

JAL cowboys in Midland, November 2, 1896. A local politician gave the outfit cigars. *Left to right, top:* Walter Cowden, Bert Rawlins, I. C. Bell, Rorie Cowden, Figure Moore. *Middle:* Fred Bartlett, Top Heard, Spencer Jowell. *Bottom:* Bi Johnson, Gene Cowden, Gene Bartlett.

Tumbleweeds

Branding 2001

May 7, 2001—My last night on the Marsh Ranch, north of the Cowdens about 40 air miles, but who's flying? The drive is 120 miles, by way of Las Vegas, New Mexico, along parts of the old Santa Fe Trail. Running through the ranch is the Mora River, flowing east out of the Sangre de Cristos, making this a dramatically different operation from Sam's. Although the Marshes summer cattle, the ranch is managed for wildlife as much as stock. Beyond the lovely, luxurious headquarters, you find wonderful country and habitants. Along the river, meadows and bosques—from the Spanish for "woods"—are home to an abundant population of deer, turkey, and quail; on open land back from the river range trophy antelope and elk. Ducks flock to a large lake; a smaller lake was built for trout fishing. This is a hunter's paradise, which is the same direction many other ranches have moved to help stay solvent. As Sam and I have often discussed, ranching increasingly takes on a demonstration quality rather than a truly commercial way of life. "Anymore it's going to be dude ranches and hunting," says Sam at his most fatalistic. "Entertainment."

This isn't unprecedented, of course; the open range was renowned for hunting wildlife as well as raising cattle. Idle European royalty and eastern adventurers engaged western celebrities to guide them in the slaughter of game; one famous buffalo hunt involved Grand Duke Alexis of Russia, Buffalo Bill, General George Custer, and other luminaries. I've got no gripe with blood sports, when practiced with some perspective and respect for other species and our common world. I love fresh cutthroat sautéed in butter by the stream, dove wrapped in bacon cooked over an open fire, wild duck simmered in wine. I've got blood on my hands, but trophies don't interest me, nor does big game much. Fly rods and shotguns, yes; high-powered rifles with scopes, nope.

Here on the Marsh Ranch, they have a small buffalo herd, likely out of the same nostalgia that led Charles Goodnight to preserve a herd on his Palo Duro ranch. The 30 million to 70 million bison estimated to have once roamed North America are down to some 350,000. Ted Turner's got a big herd on his Vermejo Park Ranch north of here. He pays lip service to their heritage and commercial potential but nevertheless feeds the "wild" beasts range cubes every day. You can kill an old bull there for a price; weekly hunts run about ten thousand dollars.

Buffalo and the domestic cattle that replaced them on western grazing lands are different in habits, handling, taste, nutrition, and other ways. Buffalo—*Bison bison*—live longer, graze on a wider variety of plant life, convert that to more-nutritional meat (much less fat, fewer calories, and fewer carbohydrates, say U.S. Department of Agriculture studies), but are more difficult to manage. A conventional barbed-wire fence doesn't discourage a buffalo bull or cow that wants out or in. "They are not domestic animals," warns all the literature. I enjoyed seeing the Marsh herd, woolly and scruffy, standing around or dusting themselves in their wallows, but a sadness inevitably hung in the air. Buffalo seem both the past and, in ways, the future of the cattle business: maintained as reminders of a vanishing way of life. As Sam says, entertainment.

Tomorrow the real work begins.

May 8, 2001—I'm in the bunkhouse, waiting for Sam to come back from showing the ranch to Steve Hoar, Jake Shanklin's friend from Alberta, Canada. Jake is Souli's younger brother, "a riding son of a gun," according to Sam. Jake and Steve are here to help with branding; both train horses for a living and will ride Jake's string of seven polo ponies. I say ponies, but they are Thoroughbred horses that Jake buys off the racetrack, then turns into polo mounts. No "pony" about them. Jake likes to bring them out here because they get lots of work during roundups. "Don't even think about getting on one," Sam warned me. "I won't ride them myself."

Unfortunately, Souli won't make it this year, and I'll miss his wit. Jerry Young, also here last year, is expected but hasn't shown up or called. So it may be Sam, Herman, Jake, Steve, me, and Geronimo, the hand Sam hired for the summer. On days we brand, neighbors will help, like always.

I need help. My back is, as they say in Mexico, *chingada*—fucked. I've had trouble with it since my first night in Roswell, and it tightened up again while I was staying in the Marsh Ranch bunkhouse. I visited a chiropractor in Santa Rosa; the "adjustment" seems to have helped some, but I'm dubious. How can chiropractic voodoo stand up to a week in the saddle and branding pens? And it's not seemly for cowboys to complain. Strong and silent, we. Ranching was, and is, a culture of profound stoicism, which arises from many things but especially the nature of the work, its place in the physical world—its seasons, cycles, elements, and events beyond human control. Weather and landscape require human for-bearance. Patience in the face of a drought or blizzard is wisdom, unless one wants to go crazy or get killed. I think of Yeats's poem with grieving Cuchulain trying to battle back the waves after his son's death, a heroic human response nevertheless ineffectual except as emotional release. The waves continue to come in, the wind continues to blow, the skies remain cloudless.

Cattle and horses are more compliant than nature but hardly perfectly so; working with them suggests patient communication to the degree that species can speak to one another. They are willing, sometimes, to cooperate, but we must make clear what we expect of them. Movement toward a set of pens in a pasture or separation of one cow from a herd of cows—such responses are possible and more so if we attempt to under-stand cattle or quarter horses, their points of view. At best, intersection of man and beast, landscape and labor, produces character like Sam's—attentive, thoughtful, hardworking, unhurried. Perspective: that's what ranching out here in the open can provide. Who said that experience is not what happens to us but what we make of what happens to us? I welcome this experience for the perspective it offers and sometimes forces upon me.

Pretty damn philosophical for a bunkhouse. The boys are back; I'd better go out to Jake's horse trailer and drink beer and swap stories.

May 9, 2001—I'm up first, since I inherited the cookstove in Jerry Young's absence. Last year Jerry would rise at 4:30, his bones stiff from brief sleep, long years, hard falls from horses. He grew up in Montana, where he rode broncs but was "too tall to win much." A bronc or bull rider should be small, with a low center of gravity, so he can stick like a tick to the back of a bucking animal. That's not Jerry, who is six-two or -three, with fused vertebrae that compelled him last year to get down on hands and knees to see what the fires on the cookstove were doing. You'd walk in and Jerry was on the floor, peering under the big griddle of the Garland, sausage and home fries sizzling on top. On mornings he was too hungover, someone else filled in—like me.

Cowboy cooks are famous in the annals of trail drives. On one drive the policy was "If you complain about the cooking, you take over." The cook was sick of cooking, but no one would grouse about his meals. So he devised a plan. He gathered a mess of cow chips—often a source of fuel on the prairie—and cooked them up into a pie. After the regular supper of sourdough biscuits, beans, and bacon, he served a big slice to the hands. The first cowboy took a bite, gagged, and exclaimed, "Why, hell, this is cow chip pie!" Then said, "Good, though."

Somebody has to cook this year, and the honor often goes to the oldest, broken-down cowhand in the bunkhouse, which would be me. And my tight back has me down on all fours just like Jerry.

We wonder where he is, and keep expecting his truck and horse trailer to roll up, but so far no show. Jerry was once in the "picture business," as a cowboy stuntman and then a horse wrangler—that's where he met Souli and got onto the Cowden Ranch bandwagon. Back in his rodeo days, he broke a leg in L.A. and, after he recovered, heard that *The Rifleman* needed someone to get bucked off a horse. "It paid $165 for the buck-off, so I did it," he told me. "If you hit your spot on the shot, but the cameraman or someone else screws up and they have to reshoot, you get paid again for another buck-off. I hit my spot three times and got paid nearly $500 for the day. Many a time I'd got bucked off for nothing. So I wound up staying on. I liked the money but not the business. Spent eleven years as James Arness's double on *Gunsmoke*. After TV westerns disappeared, I did movie work." Jerry's gone through something like five wives and now lives on a little ranch up in Colorado, traveling around to various outfits, helping with cattle work for the pleasure of riding over new or familiar country. A maverick.

In this Jerry is like many young cowboys, who may work for wages but are mostly after the thrill. Horseback in big country—that's life. In fact, anything that threatens to burden them with responsibility and stability is rejected, sooner or later. Wives, homes, regular work. So it's been for years, from adventuresome boys signing on for trail drives after the Civil War to eastern writers looking for material in the new millennium. As Roy Rogers and the Sons of the Pioneers used to sing, "I'm a happy roving cowboy."

So I've got to go, rouse these sleepy cowhands, fry the bacon and eggs, saddle up, hit the trail.

Back at the bunkhouse. We moved some cattle around. Open and late cows from Shipping pasture to Creek pasture; some heifers to the horse pasture; fifty cows and calves from Volts pasture on the west side to the trap at Alamito, where we will brand tomorrow. I'm sore today even though we didn't ride a great deal—enough to rub a little bark off my legs and stretch muscles I don't normally use.

Jake's got the same sharp wit as his brother and, like Souli, is a natural athlete. Just as the herd was almost moved into the trap today, a couple of big bull calves bolted—hightailing back where they thought their mamas were. Jake and I took off, trying to get around them and get them turned. They would not be stopped or turned or herded, so Jake broke out his rope. His racehorse was certainly fast enough but had no clue what roping was about. When Jake roped the calf, the horse broke wrong and Jake lost his rope, which the calf dragged off. He got after the calf again, jumped off his horse, caught the rope on the ground, lost it, ran down the calf on foot, wrestled him down, tied him off. While that little dance played out, I went chasing after Jake's loose horse, and the second calf disappeared back into the pasture. Good, clean cowboy fun.

Jake, Sam, and Steve eventually located, roped, and tied down the other calf, and we hauled both back to Alamito in the trailer. Jake had saved the day, though Sam wasn't about to let him off easy. "Why, Jake, did you rope that silly calf on horseback, then throw away your rope? Then run him down on foot?" Jake said, "One of us was about to drop dead."

May 10, 2001—5:30 P.M., New Mexico time, everyone in the bunkhouse resting a bit and cleaning up. We branded this morning over at Alamito, fifty calves from cows that were open the year before but Sam kept. "They were good cows," he said, "probably better than any heifers I might have replaced them with." With many ranchers, any open cow—one that doesn't test pregnant in the fall—is sold rather than carried through to the next breeding season. Sam held onto these, all relatively young and well conformed, and the decision paid off because they remained in good shape through the winter, bred back, and calved early this year. Often a cow in poor shape because of range conditions will not conceive or, if she does, conceives late in the breeding season. Her calf is born late the next spring, and it's smaller in the fall at weaning. Ideally, ranchers want cattle to conceive over a fairly compressed breeding season, so the calf crop is uniform and early. In Mississippi we kept the bulls in for sixty-plus days—cows cycle every twenty-one days on average—and I'll have to check how long Sam's breeding season runs.

So Sam got a good conception rate and early, big calves—we know, since they were hell to flank. Because of the year without calves, the cows are fat and should hold up well this spring and summer while nursing their calves, leading to another good breeding season. Maybe. On the other hand, heifers often have difficulties conceiving and calving; Sam normally calves out a hundred heifers each year as replacements for aging cows. He hopes for ten years from a cow, which first calves as a two-year-old, giving him nine calves at best. Another rule of thumb allows only seven years for each generation of breeding stock. Since heifers require more attention during and after calving season, it makes sense that a good cow— even one that failed to conceive or lost her calf—can be a better risk than an untried heifer. Plus, as Jon Means observed when I was down on the Moon Ranch, replacement heifers take three years to produce a payday: one year until they are bred, another year to produce a calf, a third year before that calf grows and goes to market.

No rancher can do without replacement heifers, of course, and they are one good way to improve your herd. Bringing in quality bulls is the best, most efficient way to improve, since one good bull breeds anywhere from twenty to forty cows. Sam buys bulls from all over but says they've never brought replacement heifers or cows onto the ranch. "Everything's been done from within, selecting our best heifers each year." This year and last, he kept only eighty heifers, largely because of range conditions. He wants to trim the overall herd but had available pastures and bulls, so he's trying an experiment this year: selling some heifers after breeding them—"bred heifers" sell at a premium. They were good heifers but had more than half-Angus or half-Hereford blood already, and Sam prefers half-and-half stock for his cows.

Rooster Cowden was a Hereford man, like Guy before him, though Guy experimented on the ranch in Midland, using Limousin and other "exotic" bulls on Hereford cows. Lots of beef breeds, each with their individual traits and advocates, are now available: Angus, Beefmaster, Brahman, Brangus, Charolais, Hereford, Limousin, Santa Gertrudis, Simmental, Shorthorn, and Longhorn—originating in Britain, France, India, Spain, Switzerland, and the good old USA. You can crossbreed anything, using natural breeding or artificial insemination, for various climates, conditions, and purposes. The British breeds still predominate out here; Sam says ranchers now favor Angus mamas, which have a number of advantages over Herefords, chiefly in what they avoid: bag problems, eye trouble, horns. Of the calves we branded this morning—twenty-three heifers, twenty-seven steers—we had to dehorn ten or twelve. With a herd of straight Herefords, that would have been 100 percent, with the trouble and danger of one more stress on the calf. Some cattlemen, like Frank Cowden, Jr., in Midland, prefer horned cows, which he says give one another more space and don't create such tight herds, especially when being worked. Loose or tight herds were a greater issue on the open range, where you had a number of different outfits' cattle mixed in one roundup; only by identifying which cow was mother to which calf could you accurately brand that calf. Horned cattle held on the open range kept their distance—Longhorns were especially good with "personal space"—and allowed cowboys on horseback to identify their calves. Now, however, it's rare to have mixed herds, and all calves are branded and marked one way. Horns don't offer much incentive, and feed-lots don't like them.

This morning I branded, a job that usually goes to Herman. Before he left, Leonard Lujan always branded, although in later years he got to shaking so bad, the brands were all over the place or blotched from his tremors. But Sam never pulled the plug on Leonard, who couldn't do much else because of his wooden leg. I suppose Sam was considering my bad back today. Herman dehorned and castrated, as well as giving me the fine points of burning the Lazy 6. I found out quickly that your iron needs to be hot—candy-apple red when you pull it from the branding pot—and you need a light, not heavy, touch. "If you press too hard," Herman told me, "the iron will slide around." Sam added, "It's hard to do right. I don't much like to brand another man's cattle; if you screw up, it's there forever." Despite the pressure, I liked the job and quickly got over any hesitation. You step up, put your boot on the calf's muzzle, burn the brand on its cheek, move promptly back to the hip, burn the brand again there. One good, hot iron should do the trick for one calf, six or eight seconds for each brand. Among the cowboys working on the ground—flanking, branding, castrating, vaccinating, dehorning, earmarking—the man with the iron has right-of-way. Everyone understands a hot iron.

Sam asked Jake, then Steve, to drag calves, while the other flanked with Geronimo, who isn't much of a cowhand. Sam also flanked. I branded, Herman castrated, and the two girls gave vaccinations. Guyito—as Herman calls little Guy—applied wound spray to steer calves or earmarked the heifers. It didn't take long to get through fifty head. Sam, Jake, Steve, Herman, and Geronimo then went to gather Eversaw and throw the cattle into the Alamito horse pasture so we can work them tomorrow morning. I dropped off cowboys and hauled some cows.

After lunch, I drove around with Sam, who had more to say about cattle breeds when we observed one of his bulls with a broken dick. "We have some every year break their dicks. It's nearly always the Angus bulls. Daddy always said that was because they don't have horns. Hereford bulls don't fight near as much. The black bulls fight all the time, get hurt lots, stifled in the back. When one bull's riding a cow, another bull will come in and hit him from behind. When they fight, too, black bulls are real bad about hitting from behind." Sam's little Corriente bulls, despite their size but because of their fighting horns, don't take any shit from other bulls.

That led Sam to talk about "downers"—"any cow that's down," he explained. "She's gotten sick or whatever; she's actually down. Lots of times there's nothing wrong with them, got hurt or something. You can't sell them now; sale barns won't take them. At least you could make dog food out of them, use the hide, you know." This got Sam off on government, regulations, environmentalists, everything that makes life difficult for cattlemen. "I tell you, these animal-rights wackos, they are really trying to put the ranchers off the land," he said. "What we're scared of is they're going to bring in that hoof-and-mouth."

Mad cow disease and the recent hoof-and-mouth (or foot-and-mouth) outbreak in Europe have become an issue not just for ranchers but for the general public. The current foot-and-mouth outbreak began in February and has led to destruction of herds of sheep, cattle, and swine in England and elsewhere. Foot-and-mouth rarely affects humans, but the public is jittery, producing a secondary effect, weakening markets in the United States. The last foot-and-mouth epidemic here was back in 1929, and all North America above Panama is considered free of the disease. No North American cattle have ever come down with mad cow. But both diseases are threats to ranchers and consumers and, according to Sam, mainly from the animal-rights wackos. "They've already threatened to. Oh yeah, PETA, the president of PETA. People for the Ethical Treatment of Animals—they're the worst ones. They've quoted him, got him in several newspapers and stuff. They want it in here, to destroy the industry." That's hard, but not impossible, to believe; the battle lines between agriculture and environmental interests are sharp. And for Sam the threat is real.

Quarantines are no guarantee, as British farmers will testify. Modern transportation methods—to say nothing of bioterrorists—now make quarantines difficult to maintain. Back in trail-driving days, herds of Longhorns coming up from south Texas were routinely stopped in Kansas and Missouri by local cowmen worried about "Texas fever." Apparently healthy Longhorns passing through would somehow infect midwestern cattle with the often-fatal disease, leading to armed confrontations, quarantines, and the eventual end of most trail drives. Texas fever was later found to derive from cattle ticks, to which host Longhorns developed partial immunity; ticks would drop off along the way and infect susceptible cattle.

Whatever the border, whenever the occasion, it was crossed: Longhorns transported themselves; imported breeds came in by ship from Europe during the nineteenth and twentieth centuries (and became subject to quarantines after outbreaks of disease); cattle from Mexico have long been threats nevertheless brought across the long border. An outbreak of foot-and-mouth in Mexico in the 1940s and 1950s was controlled by quarantines; another in Canada in 1952 was also kept out by a tight quarantine. Live cattle today move by truck for the most part, but disease can come in other ways, deliberate or accidental: one outbreak of foot-and-mouth was attributed to foreign sailors who bathed and washed their infected clothing in a cattle-watering tank near their dock in Galveston. Even now, a trucker with a dirty boot might start cattle dropping almost anywhere. No wonder Sam prefers to ship cattle out rather than bring them in.

May 11, 2001—4 A.M., I'm up and could kill these ranch dogs, barking at every coyote, deer, rabbit, or shadow within miles. I remember my dog doing the same on the farm in Mississippi. There's the distinction for you: stock farm versus ranch. Stock farms in the South and elsewhere are intensively managed, with lots of tractor work to grow the grass that grows cattle; ranches may be intensively managed, but it's more the cattle themselves, with range conditions dependent upon nature rather than nurture. You can't fertilize fifty thousand acres.

I'm sore and tired after only two days' work. I keep thinking about old-time cowhands on the trail from Texas to Kansas for two months, horseback for sixteen hours a day, sleeping on the ground in all sorts of weather. Physically tough fellows, and most were younger than I, but the contrast is still dramatic. Sam said last night that he doesn't get sore riding, never, but he rarely goes over a week or two without "forking" (straddling) a horse. He is always "legged up." So is Jake. Steve, who injured a leg and didn't ride for almost a year, admits he's sore after long rides Wednesday and yesterday.

Last night Sam and Jake and Steve talked horses, including the difference in how you train cow horses and polo ponies. Jake explained that racehorses are trained to ride like this—extending his hand flat out before him—and cow horses are trained to ride like this—making a fist—and in polo, a horse has to do both. Jake's flat hand represents a horse riding with its head out, against the bit, which horse and rider both use for balance; the fist represents a flexed neck and a horse prepared to cut with a cow rather than run flat out. I asked about my old horse's "style," which was to flex his neck yet fight to run flat out. "There's a good way to cure that," Jake said. "Sell the fucker." Which brings out a distinction I see between South and West: although there are competent trainers and riders in the South, I think the better horsemen, like Jake, come from western states, where there's more country, more room, more need, to ride.

We also talked about Buster Welch, a renowned cutting-horse trainer praised by Tom McGuane in *Some Horses*, which I'm reading. Just as I admire McGuane as a writer, McGuane admires Welch as a horse trainer, and Welch admires Foy Proctor as a rancher. Welch got his start at fourteen, working for Foy, a Midland cattleman who used to share an office in the First National Bank building with my grandfather. When asked once why he even kept an office, Guy said, "I want to sit on top of my money and make sure C. J. (the teller) doesn't lose it all." I'd love to talk to Welch about Foy and Guy, about cattle and horses.

4:35—I better begin breakfast. Ernest Copeland comes this morning, and Santa Rosa rancher Joe Tom White, which is good, since we have 150 cows to work. More hands on the ground will be a big help, since I am 75 percent and Geronimo, the seventeen-year-old living at Alamito with his young wife, is a real dud—dim or lazy or both. He's gone at the end of the month, none too soon for Sam and Herman, who just yesterday had to chew on him because he hadn't pumped water at Alamito. "Check the tank every other day, I told him," Herman said, his good humor strained by Geronimo's repeated failures. "That's his job, but he doesn't want to do it." Geronimo has spent days cutting down an old cottonwood beside Herman's house, going through a chainsaw twice and now an axe he sharpens after every few blows. The tree still stands. Geronimo's a kid and doesn't want to work hard physically, which won't do on a ranch, particularly for someone hired for that very purpose. Sam still regrets the loss of Juan last year. Juan left because his wife didn't like the isolation of Alamito, which is considerable, and more so for Mexicans who don't speak English and don't have family close. Ranching in any case is a man's world, women usually only fitfully adapting to its distances, and I can imagine a young Mexican woman fairly lost within its silences. They moved to Kansas or somewhere she had family.

On the other hand was Leonard Lujan and his family, who spent years over at Alamito, apparently happy. Rather than feel out of place, Leonard's wife, Selena, took the ranch to heart, worked to make it hers. Not just the house. She spent day after day for years, grubbing cactus from the Alamito horse pasture, a thousand acres next to the house. The project satisfied her nicely: with axe, pick, and hoe, she went from cholla to cholla, chopping them down, digging them out by the roots, making her mark. Today not a single cactus stands in that pasture—an unbroken expanse of grass that fairly sings her name. It may go by Alamito horse pasture, but I think of it as Selena's pasture.

Back from Alamito, branding the Eversaw herd of mostly black baldies. Joe Tom roped the first fifty calves, while Sam and Steve flanked on one side and Jake and Geronimo flanked on the other. "Oh thanks, Sam," Jake teased for that pairing. But Jake speaks perfect Spanish and could tell Geronimo what to do. Herman and I branded, Ernest castrated and earmarked, Kathy and Abby gave vaccinations, Guy applied wound spray to the steers, and Hannah ran irons to the branders. Jake finished dragging after Joe Tom. The work went without a hitch. Not one wreck the whole morning, the nearest being a couple of calves that threatened to run over the firepot, and would have if the draggers had been hard-and-fast, their ropes tied to the horn, rather than dallied. Joe Tom and Jake each had to drop the rope so their calves could scramble back toward the herd rather than wreck the works. Joe Tom kept us in calves at a fine pace, but once he was on the ground, things were harder for him. He has a great big belly and walks with difficulty because, Sam told me, years ago Joe Tom went to his father-in-law's to talk with his estranged wife, the conversation deteriorated, and suddenly she was shouting, "Shoot him, Daddy! Shoot him!" Daddy did just that, and now Joe Tom has a bad leg, kidney problems, a different wife. And a son who's a first-class cowboy—Kasey, who helped here last year.

I spent this afternoon reviewing material on horses, everything from mustangs to JAL ponies to modern horse whisperers. The history of cattle ranching is naturally inseparable from the history of the horse in North America; "cow boys" are more closely identified with their chargers than their charges. For anyone living in the Southwest, riding was a fact of life:

> To put the matter briefly, there arose in Europe two traditions of horsemanship, or horse culture—the one that of a settled people with whom horses were but one of the incidents of life, and the other the tradition of the nomadic people to whom horses were vital. Both traditions found their way to America, and each found its appropriate environment. (Walter Prescott Webb, *The Great Plains*)

A "plains" saying is that "a white man will abandon a horse as broken down and utterly unable to go further; a Mexican will then

mount and ride him fifty miles and abandon him; an Indian will
then mount and ride him for a week." (Lieutenant Colonel Richard
Dodge, *The Plains of the Great West*)

Dodge's comparison calls up the long history of the horse on the plains
before Anglos appeared as riders and horse breeders. Plains Indian
mustangs descended from Spanish horses first introduced to Mexico; the
Spanish horse was itself a product of Arabian stock brought to Spain by
Moorish invaders in the eighth century. In all these locations—American,
Mexican, and Spanish plains, North African deserts, the Asian steppes
before that—horses developed in relation to landscape the qualities that
made them survive and thrive. In *The Mustangs,* J. Frank Dobie praised
Arabians for "iron muscles, bones of ivory density, steel hoofs, a tail that
is the flag of its patrician heritage, a neck as lovely in arch as any curve that
Phidias ever dreamed, refined head, gazelle-like concavity between
forehead and delicate nostrils, eyes luminous with intelligence, gentleness
and spirit, and, burning steadily in every fiber, the flame of vitality."

Such virtues tended to pale as Spanish horses ran wild in the American
Southwest—free from selective breeding practiced by Spaniard and Arab,
developing by natural selection—although mustangs retained vitality if
nothing else. The best ones captured by Comanches were turned into war
ponies; the worst into meals. Marcy named Mustang Springs on the southern
Llano Estacado for horse trails leading to water there, but observations of
surveyor A. B. Gray suggest the relationship of horse and Indian: "The
name of these Springs is derived from their being the resort of *Mustangs*
for water; or, probably because it is a general camping place for the
Camanches and other Indians, in their predatory excursions to the Rio
Grande; and where the fattest animals are selected for feast occasions.
These feasts must be very frequent, as it is a perfect golgotha of horses'
skulls and bones." Farther south on the Comanche trail to and from
Mexico, Horsehead Crossing on the Pecos River also offered up skulls and
bones, from stolen Mexican horses that foundered after hard drives by
Comanche raiders. In general, mustangs demonstrated incredible hardiness,
if diminished "beauty." Dodge related the story of American soldiers
betting on their prize Kentucky horses and mares against a Comanche's
"miserable sheep of a pony, with legs like churns, a three-inch coat of
rough hair stuck out all over the body, and a general expression of neglect,
helplessness, and patient suffering"—which beast thoroughly outraced all
competition. Dodge reported that "the last fifty yards of the course was

run by the pony with the rider sitting face to his tail, making hideous grimaces, and beckoning to the rider of the mare to come on."

This racehorse in sheep's clothing had recently won its Comanche owner six hundred Kickapoo horses; Plains Indians throughout the eighteenth and nineteenth centuries—by wages and warfare, by raids of Mexican villages and Texan frontier ranches, by roundups of mustangs—accumulated great herds. Marcy noted: "The only property of these people, with the exception of a few articles belonging to their domestic economy, consists entirely in horses and mules, of which they possess great numbers. These are mostly pillaged from the Mexicans, as is evident from the brand which is found upon them. The most successful horse thieves among them own from fifty to two hundred animals."

As for riders of such beasts, I ran across this in McGuane's *Some Horses:* "Poor horsemanship consists in suggesting that man and horse are separate. A horseman afoot is a wingless broken thing, tyrannized by gravity. I have often been astounded after a great performance by horse and rider to encounter the rider afterward, a crumpled figure, negligible in every discernible way, a defeated, aging little man." That echoes what was written in *Letters and Notes on the North American Indians,* where George Catlin observed that "Camanchees are in stature rather low, and in person often approaching to corpulency. In their movements they are heavy and ungraceful; and on their feet, one of the most unattractive and slovenly-looking races of Indians that I have ever seen; but the moment they mount their horses, they seem at once metamorphosed, and surprise the spectator with the ease and elegance of their movements. A Camanchee on his feet is out of his element . . . but the moment he lays his hand upon his horse, his face even becomes handsome, and he gracefully flies away like a different being."

This recalls a visit to the ranch last summer with my kids. A couple of Brangus bulls had gotten through the fence from the Bar Y, so my son, Guy, and I rode out with Herman and Juan to drive the bulls to headquarters, where we could corral them and trailer them back to the Bar Y. Sam didn't want to simply push them back onto the Bar Y—they'd get right through the fence again, in with Sam's heifers. Hauling them to Bar Y headquarters would remove them farther from temptation and let the ranch manager know his bulls were getting out. More trouble, said Sam, but what a rancher in the West did for his neighbor.

First we had to pen the bulls. Guy had done a little riding the evening before, and some the previous summer, most of it at headquarters rather than out on the ranch, and none of it involving cattle. He was an attentive student, listening to Sam's calm, patient instructions about how to ride the right way. Riding a cow horse is entirely different from taking a trail ride on some jaded nag in a line of jaded nags, all of them "hard mouths," all trying to crop a little grass or rub their riders off against any convenient obstacle—horses with bad manners. Sam won't have a horse with bad manners, although they might have different abilities and experience. Guy was riding Hollywood, the top of Sam's string. That was an honor, evidence of Sam's habit of granting a man respect—giving him the benefit of the doubt, assuming he will do his work well.

But Guy was still a boy, just fourteen, a novice rider, no cowboy. Sam put him on Hollywood because he knew horse would take care of rider, not vice versa. All this was unspoken, a decision Sam reached and we all accepted. It was his ranch, he was *el jefe, el patrón*, the boss. Just as he put faith in them, cowboys accepted his judgment—code of the West. I rode Pajamas, Juan rode Louie, and Herman rode Streak, still recovering from some bad cuts suffered when he ran through a barbed-wire fence during a thunderstorm one night.

So it was Guy's first time working cattle. An ideal task, simple enough for success but thoroughly authentic. We were, after all, moving bulls, which always had the potential for trouble. Imagine the thrill—a kid on horseback riding with real cowboys on a ranch in New Mexico. Guy's demeanor was alert and calm at once. I could see him watch us, trying to learn, do the right thing, earn our respect and self-respect in the saddle. And wasn't he riding Hollywood? That both encouraged me and gave me pause. Guy knew to pay attention, especially once we got to the bulls, because Hollywood would know quickly what to do and he would do it. Even experienced riders can get dumped from a good cutting horse that is responding instantaneously to the moves a cow makes. Our job was straightforward, four riders and two bulls, so I expected everything would go well.

We rode to the top of a rise and spotted the bulls hanging along the east fence, where they'd been for almost twenty-four hours, since Sam and Herman had driven them into the trap. Sam's heifers were just across the fence, so the bulls weren't about to wander away, even for water. *Amor vincet omnia*—love conquers all. Herman sent Juan across the pasture to flank the bulls on one side, and he sent Guy and me down to move the bulls off the fence while he took the other flank. There was no hurry, no need to rush and excite the bulls. They were big, young bulls and, I imagine, somewhat intimidating to Guy; I was glad Brangus do not have horns. As we moved along the fence toward them, they spotted Juan holding down the far flank and turned away from the fence, west toward headquarters, just like we wanted.

When moving cattle, from the days of trail drives to the present, cowboys traditionally assume certain roles and positions, usually based on experience. At the head of the herd ride point men, who guide cattle and cowboys along; on either side cowboys ride swing, then flank, positions, turning or keeping the herd from scattering; at the rear the least experienced cowboys ride drag, pushing the herd along, eating dust. The moving herd takes the shape of an elongated, fluid triangle; all cowboys must have a common understanding of the shape and pace of movement across the landscape. On long drives from Texas to Kansas or Montana, herds grazed along so the cattle would not lose weight. At times, however, cowboys actually had to drive the herd—across a waterless stretch or when crossing rivers—and men riding drag could push the cattle too fast or let them get strung out, affecting flank riders and others up ahead. Such matters were more critical, of course, with a herd of two thousand Longhorns. All we had to do was push two bulls across one pasture, catch the fence leading back to headquarters, then pen the bulls, potentially the most difficult task.

It was a fine morning—light wind out of the west, a few red shreds of clouds at dawn—and a pleasure to ride across beautiful country, observing cattle, horses, and men. I experienced a transcendent moment, when I saw my son change from boy to man. On the ground back at headquarters, Guy was an innocent kid, but now, on the back of a horse, he was a man. He was, for all those two Brangus bulls knew, a top cowhand. Everyone showed him respect: Sam Cowden, Herman Martinez, Hollywood, Brangus bulls, his father. The critical element was his elevation to the back of a horse; on foot the bulls would have ignored him, or worse. For ages the distinction between men afoot and horsemen has been profound, riders assuming both actual and symbolic power. The literal meaning of the Spanish *peón* is "pedestrian," man afoot; *caballero* means "rider," man on horseback—a position of status to which *peones* could only aspire. Between these would eventually appear *vaqueros*, Hispanic cowboys—peones no longer, because of their mounts. And, for example, the notion of chivalry, with its evocation of knights and courts, heroic acts and honor, turns upon the French word *chevalier* and its Latin root *caballarius*, or "horseman." Mounted on Hollywood was Sir Guy.

Initially the bulls wanted to turn back, but Herman and Guy on point and flank kept them headed west; I hardly did a thing on drag. When we approached the fence line leading to headquarters, the bulls' heads picked up as they smelled water on the western breeze, and they moved quickly and easily along. The only pause came when horses in the next pasture thundered up to the fence, eager to participate even though unsaddled and riderless. The bulls balked a moment, then trotted on to the water trough at the corrals. After watering, they moved easily into the pens.

The final stage illustrated how clearly the horse makes the cowboy. Once on foot in the corrals, we were hardly formidable to the bulls, agitated and surly in the close quarters, capable in an instant of turning on any of us, sending us scrambling up a corral fence or running us over. We had some trouble getting them loaded on the trailer, and Guy again became an innocent kid, with little clue how to help and no Hollywood to show him the way. The whole process, from his ascent and heroic transformation to descent, demonstrated much about the enduring allure of cowboy life. Ten-or-twelve-year-old boys might become men on horseback, leaping over awkward lives on the ground, suddenly great warriors.

May 12, 2001—Today we gather Bogie, and Sam's a little concerned. Jake and Steve left yesterday for Roy, New Mexico, to help Ty Monisette with some steers he has to distribute. It's a chance for Jake and Steve to ride the Canadian River country, which Sam describes in glowing terms—rough, mountainous, beautiful—so it's hard to fault them for taking off. Ty was completely without cowboys to help him scatter the cattle, and Sam appreciates that problem, even if it leaves us shorthanded. We'll just have to get Bogie gathered ourselves: Sam, Herman, Geronimo, and me.

Sam's worries are with the size and lay of Bogie—seventy-six hundred acres, with a mesa spur cutting up the pasture—and with his crew. On their Thoroughbreds, what were once called long horses, Jake and Steve would have been valuable on the outside, riding far reaches of the pasture so other cowboys could cover nooks and crannies closer in. Imagine a search party and the importance of a tight net—yet we are only four riders. Though Sam has perfect confidence in Herman and reasonable faith in me, no one depends upon Geronimo. He speaks no English, but there's no language barrier: Herman will give him instructions before we separate to gather our designated areas. More important is Geronimo's limited knowledge of the landscape and cattle work; even more critical is that he's shown himself to be indifferent and sometimes sulky. Sam says, "He's got no *try* in him." And Geronimo leaves Herman—who works diligently, tirelessly, cheerfully—continually shaking his head.

Geronimo is the latest of many hands who've come to the ranch for indeterminate stays. Sometimes they last a summer, sometimes two, sometimes only a week before they decide, or Sam and Herman decide, they won't cut it. When Leonard Lujan retired a few years ago, it left Sam and Herman looking for help, especially in summer. Cowboying traditionally has been seasonal work rather than year-round employment; from calving in spring through shipping in fall there's plenty to do; then the work drops off. In the heyday of cattle drives following the Civil War, hundreds of adventurous, inexperienced young men came west to become cowboys. Most never made more than a single trail drive. What sounded grand in a dime novel or a newspaper account was sometimes demanding and dangerous but often monotonous work. Riding drag and eating dust for two months broke the spell for a boy of seventeen—Geronimo's age. Cowboys worked long hours at low pay under poor conditions; those not bred to it, or not stubbornly wedded to their own sense of freedom, rarely lasted long. Once branding is done, Geronimo is out of here, back to Mexico.

Back from gathering Bogie. We started about 7:00, the morning overcast, with sprinkles as we saddled up. We tied slickers on our saddles. "I haven't worn a slicker in ten years!" Sam said. I rode Louie, and Sam lent me some of his chinks—knee-length chaps—which turned out to be helpful in reducing chafing. I also wore a back brace but had little trouble except for the usual stiffness of the crotch and knees after much riding. I probably rode twelve to fifteen miles, start to finish. Sam took the outside on the south, Herman took the outside on the north, Geronimo was in the middle on the north, me in the middle on the south. My instructions from Sam: "Try to hit your rims and stay on top—too rough on your horse to go up and down." But lazy Geronimo didn't know the country well or the drill, so I rode hard, trying to tie in with Sam and hold the middle. "Good, though"—a favorable morning, with a cool breeze and overcast skies that didn't pour on us. Louie's nice to ride, a smooth fast trot and a slow lope that make it easy to cover country without getting beat to hell. He's over fifteen years old but held up fine, until he got a little tired and slow at the end. I rode well enough, much more capable and comfortable than the first time out.

The cattle were mostly Herefords, including the oldest on the ranch, so they'd been worked plenty of times. Sam expected they'd be balky, but the gather went well. I hooked up with Sam at just the right time to keep the cattle he'd thrown off the mesa and out of the draws from scattering back to their old haunts. We four were able to successfully sweep the pasture—didn't miss a single head, according to Herman's count. As often happens in gathering Bogie, two herds accumulated, one on the north, another on the south, and we met up right at Poso windmill, there by 11:30. We pushed the cattle into Shipping pasture, let them settle, and rode back to the house.

Now I'm trying to make horse sense. A couple of passages on mustangs—a term derived from the Spanish *mesteño* or *monstenco*, "wild" or "stray"—from Catlin's *Letters and Notes on the North American Indians:*

> Some were milk-white, some jet black—others were sorrel, and bay, and dream colour—many were of an iron grey; and others were pied, containing a variety of colours on the same animal. Their manes were very profuse, and hanging in the wildest confusion over their necks and faces—and their long tails swept the ground.
>
> The wild horse of these regions is a small, but very powerful animal; with an exceedingly prominent eye, sharp nose, high nostril, small feet and delicate leg; and undoubtedly, have sprung from a stock introduced by the Spaniards, at the time of the invasion of Mexico; which having strayed off upon the prairies, have run wild, and stocked the plains. . . . This useful animal has been of great service to the Indians living on these vast plains, enabling them to take their game more easily, to carry their burthens, etc.; and no doubt, render them better and handier service than if they were of a larger and heavier breed.

This morning Geronimo was mounted on a tough little grulla, smoky blue—a Mexican horse that echoes Catlin's description. Sam bought Macho from Souli with the idea he might make a kid horse. No way. Macho's the right size but "humpy" every morning, offering to buck with Geronimo. That feral spirit has advantages, since Macho never quits; Geronimo's ridden him every day for a week. Sam was riding another *cuñado* (Spanish for "brother-in-law," with connotations of getting screwed in a deal) special. Papas is a skinny bay, not so feisty as Macho but no angel. *Papas* means "potatoes" in Spanish, and Papas Calientes—Hot Potatoes— is more like it. "Souli got me good on these two," says Sam good-naturedly. "But Souli always rides colts or outlaws, and he probably doesn't think anything of it."

Like any sensible rider, Sam prefers a well-trained, or at least teachable, cow horse. Most are American quarter horse stock, although few are registered. Quarter horses, so-called for their speed over a quarter-mile racecourse, were developed from Arabian and Thoroughbred bloodlines in colonial America and particularly on southwestern ranges. Just as Longhorns were crossed with "improved" British or other European cattle, native horses were transformed by English Thoroughbred blood. Now a recognized, distinct breed, quarter horses are more compact and heavily muscled than Thoroughbreds, quicker at the start, easier to handle, more suited to working cattle, which requires horse and rider to react rather than simply run. Yet quarter horses are still valued as racehorses; the All American Futurity at Ruidoso Downs, New Mexico, offers a total purse of two million dollars. It takes all of twenty seconds to run the quarter-mile.

More important to Sam than raw speed are cow sense and handling, since all his horses are "using" stock first and foremost. He may like team roping but makes his living ranching, not at rodeos. Most time horseback is spent gathering, sorting, and moving cattle. Branding as practiced on this ranch calls for roping and dragging calves—some places simply run their calves down a chute and work them on a calf table—but roping is not a daily routine. Sam's horses all learn to watch cattle, heading them back to the herd if they stray, cutting them out when required. I noticed while riding Hollywood how deliberately he moves, sorts, or holds a herd. However calm he appears, you'd better be alert, because if a critter bolts, Hollywood is instantly in motion. Of all Sam's string—including Pajamas, Dunnie, Louie, Red, Streak, Crow, Smoky, Papas, and Macho—Hollywood has the best AQHA (American Quarter Horse Association) pedigree.

The Cowdens have not typically sprung for fancy horse stock, although they've had good horses over the generations. "You have always heard it said that there was not any money in horses," wrote W. C. Cochran of the JAL days. "When the Cowdens left Palo Pinto they had seven $25 Spanish mares that they rode for cow ponies. They rode them until they wore them out, then turned them loose on the range to raise horses. They never bought a horse, but raised all their horses from these seven mares that they made their big fortunes on. They sold $15,000 worth of polo horses, and had $40,000 worth when they sold out."

This sounds a lot like Sam's approach to his cattle herd, but he routinely buys horses. "Daddy told me I'd always be looking for horses," says Sam. "He was right. Especially with kids. Their needs change, horses get hurt, get old. And anymore, with team roping and other rodeo events, it's hard to find horses. They've gotten expensive." No more twenty-five-dollar mares—average ranch horses cost a hundred times that. "A really good horse can run five thousand dollars," Sam says. "I don't know who's spending that kind of money. I've never bought a made ranch horse. We always bought colts or prospects, ones that were started. We finished training them. Just a prospect, a young horse, costs two thousand dollars."

So there are always new and experienced horses in the mix, each requiring different handling. Sam knows what to do; he's been riding since he could walk, like the Comanches that Dodge described: "Riding is second nature to him. Strapped astride of a horse when scarcely able to walk, he does not, when a man, remember a time when he could not ride. . . . He is too nearly a Centaur to be surpassed by any." Centaur Sam Cowden says, "I went with Daddy when I was four. We all rode. Patty was the real natural. I remember one time when she was little, she walked behind a stallion, a stud horse, and as she passed by, flipped up his tail. Daddy just about had a fit. She was absolutely fearless." Patty went on to win a college rodeo scholarship, and Sam spent all his free time in college "drinking beer and roping." He recalls, "Like kids today shoot hoops, we'd throw trick loops. 'Figure eight over both horns.' 'Hoolihan, right horn.' I was *country*."

Did he have a choice? Sure, but Sam was never going to choose anything but ranch life.

May 14, 2001—The cowboys are out at Jake's trailer, drinking beer, talking shop—horses and horseshoes; saddles, bridles, and bits; chaps and boots; ropes and knots and knives—while they practice the cowboy arts. Jake is shaping old horseshoes so they fit down over a horse's ankle, which effectively stops them from pawing. He had anvil and hammer in his trailer, along with myriad halters, bridles, saddles, folding chairs, and other items. Ty Monisette is cutting off eight feet from a forty-three-foot poly rope— as opposed to nylon or grass—that Jake had found too long and cumbersome, requiring too many loops in his hand. Ty cut the rope, wrapped tape around it a foot from his cut, unwrapped the three strands, and tied a knot in the end to keep it from unraveling. The knot used is either a Theodore or a Turk's head. The scene resembles descriptions of sailors' lives aboard ship, downtime spent carving scrimshaw and such, keeping boredom in check.

Cutting back a new rope led to more talk about differences between Texas cowboys and buckaroos. Sam gives credit to the buckaroos on the Bar Y and the Conchas Ranch for gentling down their cattle considerably. Ty worked awhile for the Bar Y and doesn't always compliment their practices. For example, they don't worry if their cattle get over onto Sam's ranch; often they don't even know, because the operation is so big and loosely tended. Just today we had haul a cow and calf back to the Bar Y;

she had gotten into Conchas pasture from the north with her calf. She was a half-crazy Brahman cross; in the pens we cut her out first thing and loaded her into Sam's covered trailer to await transport. She had one horn pointing up, and one forward, like some dangerous redneck with bad teeth. By the time she was unloaded at the Bar Y, she had carved a rip in the canvas top over the metal cage of the trailer. Good riddance.

The branding crew was Sam, Herman, me, Geronimo, Jake, Steve, Pete Marez, Ty Monisette, Kathy, and the kids, who spent only half the time in the pens since Jake was riding a green and "silly" horse—somewhat dangerous for everyone involved. When we were sorting, Sam cut out the dries and heavies where Jake happened to be stationed on his nutty horse—actually just ignorant and wound up, coming to the ranch straight from the racetrack. Not the situation Sam would pick, but he didn't want to say anything. And Jake does so well with his horses that nothing went wrong, either on the sort or in the pens. The mare quieted down after dragging a few and began to get the hang of things. We were all on our toes, however, so we wouldn't get kicked or run over.

Steve dragged first on his green horse, just twenty calves because of a bad arm. The other day while flanking calves, he felt something rip; last night his whole bicep was a nasty shade of violet. Today after roping, he branded, another thoughtful use of wounded hands by Sam. Ty followed Steve dragging calves, Jake followed Ty, and Pete finished up. It went a bit slow because of the green horses, but everything else was fine. Tomorrow is our last day, branding in the Shipping pens.

May 15, 2001—Another branding bites the dust. Today Sam's neighbor Bill Mitchell came to help, along with Sam, Herman, Jake, Steve, Ty, me, and Geronimo.

We branded at the Shipping pens, and everything went perfectly. In fact, the whole week went like clockwork . . . until the last minute. We came in, washed up, and sat down on the front porch of the ranch house for a big, satisfying meal. Everyone was finishing their pie and chewing on toothpicks when Geronimo called Herman to the back. Geronimo had been invited to eat but declined to join us. Herman returned, and then Sam went back with him. About then we saw Geronimo's truck race away, back to Alamito. Something was up.

Herman came back, shaking his head. *Geronimo!* Our young cowhand had somehow backed his truck *through the wall of the bunkhouse*. We all walked out to see the damage. Sure enough, where the window to a bedroom used to be was a gaping hole of broken cinder blocks and shattered wood. We stared into the disheveled room, floor covered with rubble, bed dusted with glass. A low wall outside the back door had been clipped in the process. "How did he do it?" we all wondered, and no one could answer. Herman didn't know, even though he'd talked to Geronimo, who claimed the truck had rolled from where it was parked, uphill near the saddle barn. Trouble was, Bill Mitchell's truck and trailer blocked that path, so Geronimo's runaway truck had to swing around the trailer to smash the bunkhouse. "I think he was pissed off," Herman said. "I don't know if he did it on purpose, but I think he was pissed off." "Oh, it was deliberate," said Jake. "Had to be." All week long everyone tried unsuccessfully to get Geronimo to flank calves right so we could get the work done. Geronimo didn't give a shit and eventually seemed to slack off and screw up on purpose, letting calves flail around so it was hard to brand, vaccinate, castrate. Sam and Herman had both lost patience with him but wanted to get through branding before they ran him off the ranch. Even now, to Sam's credit, he stood before the wreckage of the bunkhouse, shaking his head but smiling.

"And I was just thinking, Sam," I said, "how we'd gone all week long without a single wreck."

"And now we get a real one," Sam said.

It was true: the week had given us fair weather and good work, despite the weak link of Geronimo. Now the link had snapped.

"No insurance?" I asked Sam.

"Insurance? No, sir. Not out here."

Geronimo was gone this afternoon, with his *esposa*. Herman called him at Alamito, told him to get everything together and el patrón would take him into town. "But I am staying to the end of the month," he tried to say. No, he was not.

Sam had him back at his uncle's house in Santa Rosa before dark. *¡Adiós, Geronimo!*

Yo también. Me too—and Jake, Steve, Ty. All of us off in the morning, back home to Massachusetts, Texas, Calgary. The spring stampede is over.

Home on the Range

Living at best in dugouts, cabins, and tents, and often day and night in the open air, enduring hardship, privation, and deadly dangers—[he] determined to settle down and seek the comforts and quiet repose of a good home, and to bring around himself those tender endearments without which wealth and life itself is but a blank and a failure.

Joseph McCoy, Historic Sketches of the Cattle Trade of the West and Southwest

People thought that we, that I, wouldn't make it. The bets were probably on. Just because I was a little town girl. At Alamito when we first got married, we had that old wind charger to begin with—no electricity. Gas stove, gas refrigerator. No phone, no neighbors. It was a different world; it really was. Mushy said to me, "Don't milk the cow." Don't milk the cow, because if he goes to town for something, he'll come home to milk that cow. That's what mothers told their daughters if they married a farmer or a rancher. He didn't want to ruin that cow."

Jean Cowden

Sam and Kathy Cowden, Rooster and Jean, Guy and Mushy, Big Daddy and Moner, Bill and Carrie Cowden: the roll call always in pairs, a family always in place on ranches outside Santa Rosa, Midland, Palo Pinto. Whatever location, the way of life was shared and stable, unlike the Sons of the Pioneers song "Roving Cowboy" or Patsy Montana's romantic "I Want to Be a Cowboy's Sweetheart." The Cowdens were not cowboys and their sweethearts, but settled ranchers; not single, but married men and women and children. Life was no Hollywood romance, no Roy Rogers and Dale Evans duet on Trigger and Buttermilk. The usual dramas between men and women, parents and children, families and the outside world, all apply to ranching. Nevertheless, labor and location affect everything from waking to sleep: shelter, food, clothing, education, society, culture. Even dreams. William H. Cowden dreamed of maverick Longhorns, Cowden women those first years on the JAL Range dreamed of ranch houses to replace dugouts or tents or wagon beds, and Sam dreams of flowing water. Children dreamed Indians, fast horses, dances, other worlds beyond the open horizon. They all dreamed the future as much as the past.

The actual pulled them back and engaged their energies. Meals to cook, cattle to herd, droughts to weather, wells to dig, schooling—practical matters demanding their full attention. Kathy Cowden wakes every morning at five to get her day in order; she's got lesson plans and preparations, breakfast, kids to rouse. "Good morning, Sunshine," I greet her when I visit from the bunkhouse. She shoots me a mock frown, a real smile. All the Cowdens are resolutely yet truly cheerful in the morning. Sam explains, "Daddy used to tell us, 'You be up, at the table, in good spirits.'" No moody mornings here: everyone takes daybreak in stride, like Navajos whose traditions demand they greet sunrise from their east-facing front doors, lest the sun mistake them for dead.

As life quickens inside the ranch house, Sam and Herman start their day, feeding horses, catching their mounts, making plans to move or feed cattle, check waterings, repair fences or equipment, grade roads. Sam can never be completely sure what might happen, so Kathy usually prefers to steer clear and concentrate on her responsibilities at home. Many modern ranch wives have jobs in town, for income but also for exposure and stimulation they can't get otherwise. But the Cowden Ranch is a good forty-five-minute drive over bad roads to Santa Rosa, where opportunities are limited anyway—not much of an option. Kathy likes homeschooling, and the Cowden children, still too young to yearn for a more varied social life, seem content as well. Once school is out, they play around headquarters or roam the mesa with complete freedom and satisfaction, like Sam and siblings before them. No bored hours in front of the TV. Rather, it's horses and dogs and cats, romping on the trampoline, scrambling over the haystack, riding bikes or the four-wheeler, feeding deer that now come regularly to headquarters. They have piano lessons, library and 4-H programs in Santa Rosa, trips to Albuquerque. Their mother is an organized and exacting teacher, providing structured study, but she's prepared to ditch the routine and take advantage of the ranch as a classroom, with its own materials and schedule. "Field trips," Kathy says. "I'm a *big* fan of field trips."

One obvious drawback of homeschooling on the ranch is exposure: underexposure to other children and activities, and consequent overexposure of mother/teacher with children/students. Kathy admits she can't do everything as a teacher, mother, and ranch wife, and, understandably, she can get fried toward the end of a day, however companionable the Cowden kids and Sam are. Where families in town can depend on friends and neighbors for relief, here, "widely separated from its nearest neighbor, the (ranch) has to be self-sufficient in all things." Anyone living on a ranch better develop "inner resources."

Sam—never one to stand by and watch someone else work—does his part with the kids. Horses and riding for one, but also exploring the physical world of the ranch: outings to see prehistoric Indian "writings," hunt for arrowheads, or view ruins of Spanish sheepherders' houses and corrals. And he assumes his domestic duties, often cooking on the grill and helping with shopping—a big change from the previous generation. "Rooster hardly *ever* went to town," says Sam. "Once he went in to the grocery and tried to charge a few things—they didn't know who he was! I buy groceries all the time." Sam and Kathy seem like typical modern parents, trying to negotiate gender and family roles in a busy, changing culture even as they hold onto the traditions of a ranch family. "Most people aren't like us anymore," Sam says. "We sit down and eat together. All our meals. And the kids don't get up until they've been excused. Little kids come out to visit, and it shocks the heck out of them. It's a whole lot different. We're the only people I know who have that structured life." It makes an enormous difference, certainly, when children have mother and father ever present as guides, and the Cowden kids interact well with adults—respectful yet relaxed and expressive. It's fortunate for everyone that Sam and Kathy recognize they must balance guidance with freedom and let character develop—after all, the code of the West reveres individualism. The Cowden kids are good but hardly flawless; the same goes for Sam and Kathy. For anyone in the family, we might adapt the old cowboy wisdom I quoted to Souli: "Some men's wives are angels; others are still alive."

Also, western ranch women—unlike stereotypes of their demure Southern ancestors—are famous for "spunk." I don't mean Calamity Jane or Belle Star. I mean unsung women of ability and grit, which compensate for deprivations natural to ranch life: Carrie Cowden raising twelve frontier children, Annie Mae pulling her husband out of a well, Olive leaping from a runaway wagon with infant in arms, Jean pouring out cattle feed, Kathy homeschooling three kids. Where do these women come from? Sam says, "I married *up*," but David Dary notes in *Cowboy Culture* that most "wives came from the same social stratum as their husbands. This contributed to the stability of ranch marriages, in which family life became the center of social and economic activities."

Kathy is from a ranching background, before she became a school-teacher, then married Sam. She knew what she was getting into, as much as anyone does who marries and moves to a new place. Not so for Jean Walpole, whose family came from Iowa to New Mexico. When she married Rooster in 1952, Jean had no clue; she underwent the same transition as Annie Mae, another town girl, did when marrying Guy in 1922. Annie Mae's father, Tom Patterson, owned Everybody's, a dry goods store in Midland, and he served on the school board. Annie Mae went to Midland public schools, while Guy Cowden had private schooling; they didn't meet until older. Guy fell hard and went to wooing her, despite the reservations of the Pattersons—Guy was from a good family but was somewhat of a rake. The romance came to a boil after Annie Mae was selected to represent Midland at a pageant; accompanied by a chaperone, she took the train to San Angelo. That prompted Guy—who knew who she might meet?—so he followed on the next train and persuaded her to elope. The distressed chaperone had to notify the Pattersons that their daughter had just run off to marry Guy Cowden. Midland was scandalized—MISS MIDLAND ELOPES AT PAGEANT!—but Guy and Mushy were starting an adventure of almost seventy years together, on ranches and in town.

Guy worked first for an uncle, then leased Rountree from cousin Dick Cowden in 1926. Two years later he persuaded his father to buy the ranch, at which point he leased a neighboring spread, acquired with Rube Evans from another relative. From 1928 to 1944, Guy based his cattle business fifty miles west of Midland on parts of the old JAL Range. The children were all born in town, but everyone lived on the ranch until 1937, when the eldest daughter started high school and the youngest began first grade. The Cowdens were following the custom of cattle country, as Robert Rankin recalled: "In that day and time, nearly all those early-day ranchers had a home in Midland and on the ranch. They'd spend most of their time on the ranch, but when they'd get caught up on their work, they would come back to town." Bob Beverly remembered, "Midland was head-quarters—a great cowtown at one time. They started it as a school town, a place to move their families to go to school. First time I ever saw old man Bill Cowden, he had a tent boxed up on the north side of town and was living in it, sending kids to school. At first he and Mrs. Cowden would come in with covered wagons, an old milk can or two, and camp through winter to send kids to school. They moved back to the ranch in spring."

So the Cowdens and other ranchers had dual, sometimes conflicting, allegiances: open ranges that cattle required and towns that offered families a wider world, even if they seemed to restrict the freedom and independence that characterized ranch life. Where cowboys might blow off accumulated steam and wages in town, cattlemen more often simply shifted responsibilities to business and social matters. No wonder ranchers could be hard to part from their herds, and civilized life was routinely the province of wives and children, rather than of cattlemen themselves. The men might spend weeks and months away from their families; many began ranching as bachelors and, in effect, reverted to that condition as children required schools and wives sought company in town. "For the ranchers' wives," Dary wrote, "there was little female companionship. Few hired hands had wives. Most ranches were many miles apart. Unless a rancher and his wife had one or more daughters, there was little 'woman's talk' except on infrequent visits to other ranches or to town. The towns were only county-seat villages with a few hundred people, but they provided a meeting place for ranch women." Olive Cowden tells of family gatherings at Moner and Big Daddy's home in Midland, where visiting women would all sleep on the second floor. "We were all up there one night late, talking and talking, and finally Mr. Cowden came upstairs and said, 'Girls, either speak loud enough so I can hear you or shut up.' Oh, we had fun."

Early Midland was civilized but hardly luxurious, Rankin remembered: "There was no water supply, no sewer supply; outdoor toilets they had. Every house had a windmill or had to buy water from somebody who got it out of a windmill. Wooden frames, and had about a twenty- to thirty-foot tower, and an eight- or ten-foot-diameter mill, old Eclipse wooden ones." Over time Midland became more prosperous and proper: schools were accompanied by churches as social institutions, sometimes in conflict with cowboy ways. Bob Beverly recalled a temperance battle at the turn of the century: "They had the Legal Tender saloon there, and that's where we all watered out at. Had a good hitch rack in front where we could tie our horses, and it was the principal watering place of all the range hands. . . . Midland was then a very religious town, and quite a school town—and they had big speakers there from everywhere." Beverly and other cowboys from the Quien Sabe Ranch south of Midland were passing through on a trail drive and voted against prohibition, postponing the eventual victory of temperance elements for a couple of years.

With schools, churches, banks, stores, and the railroad, Midland was the heart of West Texas and southeastern New Mexico culture and cattle business, yet ranches themselves were the fundamental units of life, more demanding and distinctive than anything found in town. "I loved it. See, it was new to me," says Olive Cowden of moving to her father-in-law's ranch as a young bride. She had attended school in El Paso and Austin, studying piano—hardly preparation for ranch life. "Moner and Big Dad, when Al and I married, got a piano that they had and put it on a wagon and sent it to the ranch, over those sand hills. I could not believe it. But the sad thing was that I could only play classical music. Well, can you imagine cowboys wanting to hear classical music? So I didn't play the piano much." Olive eventually returned to "civilized" society: "Finally I went with Mrs. Cowden. She begged me to come into town and be with her, you know. I wanted to stay out there. I liked it out there."

Like previous generations, Gene Cowden and his young sons—Guy and Al—cooperated in raising cattle, and their families met to dispel isolation, recalls Olive: "We visited back and forth. You know, we used to meet at the sand, with picnics. And oh, we had a good time. Have a big cookout. The men would cook and we'd take cake—you know, good stuff to go with it. See it was eleven miles between us. Sand. Bad sand. Sometimes we could go in the wagon. Sometimes we could go in the car, if the sand would be wet [and thus hard], but that was few and far between." One visit included telling details: "Annie Mae and Guy had us visiting. It was just a little house like we all had in those days. All the children were little. Al was lying on the daybed, close to the front door. I was standing in the kitchen door into the living room, drying a dish. I happened to look up, out the screen door, into the front yard, and Emma Sue—it was a little child of ours—she had this snake up in her arms! Well, my mouth was moving, but no sound was coming out. I was paralyzed; Annie Mae was too. Al was sitting there looking at me, thinking, "What is going on?" Finally he jumped up and went to the door and then went on out and got a hoe and killed the snake."

"You know the sad thing out there," Olive says, her lament rare and muted, "Annie Mae and I both always wanted a pretty house. That goes with a woman. But we did real well. Annie Mae could paint good. I didn't know how and I didn't want to know how. I didn't mind doing anything else. Anyway, Annie Mae could paint and have the furniture looking nice. I'd start to paint, and I'd play out pretty quick. We wanted to make it pretty, and we didn't have much to make it pretty with." As Agnes Morley Cleaveland wrote in *No Life for a Lady*, "With the cattle business the sole industry, and a working schedule of twenty-four hours a day necessary to conduct it, there was little consideration given to the comforts of daily life and practically none to the refinements. Women had to get along with what they had, which was precious little."

Occasional visits, picnics in the sand, spare homes—West Texas ranch life during the Great Depression was an isolated, stripped-down affair. Much of the time no one was around but family. My mother told me of one dramatic exception, when Guy hired an itinerant hand, "someone passing through, you know, hungry, looking for work," she recalled. "Well, this cowboy stayed a while and took a shine to Mushy—he told Guy he was going to take her off. Can you imagine? Guy said, 'Like hell you are,' and they had a big fight, right in front of the house. Fistfight. Guy whipped him and ran him off the ranch." Such bad behavior violated the code of the West, as David Dary has pointed out: "As ranches developed and women arrived—most were the wives and daughters of ranchers—new rules developed. For instance, cowboys could show no interest in the women of the household, or even an appreciation for their hospitality except by eating heartily at their table. A cowboy was to respect good decent women and protect their character."

A good decent woman like Annie Mae or Olive was clearly devoted to family and home, to traditional values and roles. To be wife and mother, then and now, usually placed ranch women inside, not always happily, as Agnes Morley Cleaveland observed: "It was this deadly staying at home month in and month out, keeping a place of refuge ready for their men when they returned from their farings-forth, that called for the greater courage, I think. Men walked in a sort of perpetual adventure, but women waited—until perhaps the lightning struck." Dorothy Ross seconded the complaint in *Stranger to the Desert*: "Worst of all to me was the monotony of unaccustomed housework. It was often too heavy for me, and was always an uninteresting drudgery. No one can deny that the men worked hard too. But they were out of doors, usually on horseback and working with animals, and they were free to meet others interested in the same business. . . . It was a truly surprised and unhappy rancher who said: 'I can't figure out why my wife went crazy. Why, she ain't been out of the kitchen in twenty years!'" Newly married Gladys Cowden was advised that the way to manage her ranching husband was "Feed the brute." She later observed, "Frank looked so glamorous to me, and I just couldn't imagine anyone connecting him to that kind of statement. Well, after I had gone on the ranch with him, and would see him sitting there and just stuffing frijole beans and onions in his spurs and cowboy hat, I thought of it many times."

Olive Cowden accepted her role, although new to her. "You know, I never had cooked," she says. "I didn't know a darned thing about cooking. A recipe would say soda, and I'd say what kind of soda? What kind of milk do you use? But I had to learn about soda and baking powder and finally did. The cowboys would come in and they'd say, 'Mrs. Cowden, if you'll bake a cake, we'll wash the dishes.' I'd say OK. So we had cake, and that pleased them." Olive's father-in-law, who learned to cook during his bachelor years on the JAL, was her first instructor. Gene Cowden also kept a garden. "A great, big garden he had," Olive remembers. "Al hated to fool with it. But Big Dad didn't care who hated it. He knew we needed a garden, and he liked it anyways. We had a lovely garden." Cowman, cook, gardener—E. P. Cowden understood the necessities of ranch life and helped convey those to the next generation, male or female. The wisdom of ranching applies not simply to working cattle but to living a unified life in a unique place.

It's not a land and life suited to everyone. "The cowboy of the west worked in a land that seemed to be grieving over something—a kind of sadness, loneliness in a deathly quiet," wrote Bob Beverly. "One not acquainted with the plains could not understand what effect it had on the mind. It produced a heartache and a sense of exile." May Price Mosley refined that sentiment: "The monotony of the prairies and the sameness of the days were either beyond endurance or tranquilly serene, according to the temperament of the person. If one were of a lonely or pessimistic nature—or dependent on others for entertainment, he usually did not stay. But if he were of a contented, naturally happy disposition, there was certainly little to disturb such content or interrupt his happiness. Occasionally one lost the day of the week; some rode miles to be set right again, while others made no pretense at knowing the day or date—nor cared." These days on the Cowden Ranch—with satellite television, Internet access, all of the technology that keeps people both in touch and apart—life still seems removed. Look around: not another structure in sight, only plains and skies that seem infinite. No laptop or virtual community fills the actual void.

With so few social institutions available to them, so little "society" at all, the Cowdens entertain themselves, as they always have. They belong to the Methodist church in Santa Rosa—a congregation dwindling for decades, as the population steadily becomes more Hispanic and Catholic—and the kids are in 4-H, which often gets them into town, but the family seems closer in spirit to the nineteenth-century Palo Pinto Cowdens than to the twentieth-century Cowdens who drew extensively upon Midland. Far from any appealing town and close to his work, Sam resembles those frontier ranchers who did little but labor, as Gil Hinshaw described: "Work, skill, endurance, enjoyment and existence were very nearly the same thing. One worked to live, lived to work and the competition in demonstrating skill—

to see who was the best—was a form of satisfaction, enjoyment. Over all, like a glass dome placed to preserve a miniature life form, was the awe-inspiring isolation." Though Sam hunts on occasion and enjoys team roping—common diversions—ranch work, with its singular demands and satisfactions, most absorbs and fulfills him: "I couldn't do anything else. I get frustrated too, even around other ranch people we know—they don't know the things Rooster taught me. I don't know of anybody else, I don't know of another soul who knows those things." Kathy adds, "He's so dedicated to this life. I think you'd just have to bury him if you took him off of here." She and the kids get away more often—to visit family in Texas or friends in Minnesota or just to make a run to Albuquerque—but have the ranch fixed firmly in the center of the universe. "I thought, whenever she came out, 'Oh, she's going to have a hard time. She's going to want to go to town every day,'" says Sam. "Now I end up buying most of the groceries because I can't get her to go to town! She's here and she likes it here." It's much the same for Herman and Debbie. Sometimes after the work ends at noon on Saturday, Herman will go in to see his mother, have a meal, visit with friends. "But if he doesn't," says Sam, "watch out. By Monday morning he is itching to get back to work. Heck, there's nothing else to do!"

Rodeo now ranges from televised national finals before big crowds in Las Vegas to winner-take-all jackpots at small-town arenas—more than two thousand rodeos annually in the United States. As it evolved from competition between cowboys on ranches—Who could ride that rank bronc? How many calves could you rope without missing?—into later contests at cow towns, where trail herds might wait weeks for sale or shipment, rodeo kept its connection to actual ranch work. One of the first organized rodeos was held in Pecos, Texas, on July 4, 1883 and offered a cash prize of forty dollars, with a reported one thousand spectators watching ranch cowboys from West Texas and New Mexico. The prize money is better these days, for Professional Rodeo Cowboys Association (PRCA) events like saddle bronc riding, bareback riding, bull riding, calf roping, team roping, and bulldogging. Cowgirls have their own organization and add big-money barrel racing to their rodeo mix. And don't forget brave, athletic rodeo clowns. Wild-cow milking! Wild-horse races! Trick riding!

Although "rough stock" events attract the most attention—bull riding in particular makes great, dangerous theater—no working cowboy ever tried to buck out a bull. Ranch horses do buck, as Agnes Morley Cleaveland remembered: "Seeing to it that the other fellow's horse pitches is routine on a frosty morning. Horses' humor is no better than that of their human brethren on cold mornings, and any unexpected happening will set them off. So yell and throw your hat under the feet of your neighbor's horse just as he mounts, and the chances are you'll have a nice little rodeo right then and there. It is not difficult to understand why we had so few gentle horses. We had so few gentle people!" Roping events also have direct parallels on ranches, as when branding or doctoring stock; many cowboys see roping as the best measure of their abilities. Rooster naturally asked, when receiving religious instruction as a child, "Could Jesus catch a calf on the first loop?" Roping events unite all the principals of ranching—man, horse, and cow— with differing demands. Where calf roping requires single contestants to rope, dismount, flank calves, and tie them down, team roping is cooperative and more dependent upon skill than athletic ability. Available to women as well as men, young and old, fit or flabby—anyone who can ride well and throw a rope—team roping has become very popular, "the fastest growing equine sport in the country," according to the U.S. Team Roping Championships. The USTRC produces amateur and open ropings, maintains the handicaps of more than 100,000 ropers (Sam is a three, of a best-possible ten), and gives out prize money—sixteen million dollars in 2000. Big time. The USTRC motto: "Steer into the New Millennium!"

Sam has been roping all his life, for work and pleasure and now makes team roping his main diversion. The team includes a "header," who throws the first loop, catching the steer by the horns or neck before the "heeler" attempts to rope both back heels. Once both ropes are on, stretched taut, riders facing one another, their time is recorded, with penalties for starting early or catching only one heel. Of the two loops, heeling is generally more difficult, although no more important; ropers tend to specialize so they can master certain skills. Sam's a heeler, and a few years back bought Buford, a mechanical steer made by Rope-O-Matic in Marathon, Texas, so he could practice heeling out on the ranch. "Just saddle up, flip the switch, and go rope!" says Rope-O-Matic. "Buford doesn't get tired" as he circles at different speeds, back legs kicking as he goes, Sam on horseback on his tail—a merry-go-round of virtual steer, real pony, real cowboy. The scene lacks only bouncy organ music to be complete. Buford saves the expense and trouble of maintaining live roping steers, and since no one's available to head a live calf—Herman doesn't rope, nor Kathy, nor any of the kids yet—Buford lets Sam practice solo. Of course, there's often no time and, after a long day in the saddle, no inclination to practice. But if a roping is coming up in Santa Rosa or beyond, Sam might give his horse and himself a refresher. Round and round go Buford, Hollywood, and Sam, his rope swinging above his head. "It's technique and timing more than anything," says Sam. "It comes back."

Sam's frequent partner is Leigh Ann Marez, and I had a chance to watch them compete at Super Loopers, the USTRC southwestern regional finals at the state fairgrounds in Albuquerque. The parking lots were full of pickups and horse trailers, riders were everywhere, merchants sold ropes and saddles and western wear, and two big arenas presented nonstop action all day. The stands were maybe half full, mostly with ropers' families and friends; I sat with Kathy and the kids waiting for Sam and Leigh Ann to come up in their rotation. A steady procession of roping teams, riders looking sharp in cowboy hats and pearl-snap shirts—no caps or T-shirts allowed—fed into the arena, eventually moving to the starting gate, positioning their horses on either side of the chute, readying their ropes, nodding slightly, then breaking at the bell. The steer races up the arena, riders in pursuit, header roping and dallying and turning the steer, heeler roping and dallying, team stretching out their ropes, and time is called. Unless someone misses a loop, of course, so the team gets a score of "no time" and can forget any payday. A good run is six to eight seconds, although pros regularly score five-second times. Sam and Leigh Ann did extremely well the first two rounds, leading all teams in their classification with a score of seventeen-point-something seconds. They slipped a little in the third round, then missed a loop in the fourth, which put them out of the money and meant there'd be no trip to the national team roping finals. No one showed great disappointment, since they had spent an exciting weekend, had seen friends from all over, and were ready to get back to the ranch. Pro rodeo life—with road trips from Albuquerque to Phoenix to Denver to Cheyenne to Calgary to Pendleton to Kalispell to Tulsa to wherever whenever, its endless homeless roving—is not the Cowden way.

What's more likely, other than occasional ropings in Santa Rosa, is some visitor to the ranch now and then, someone like me. Traditionally, a few actual cowboys moved from ranch to ranch, "riding the grub line" when no work was available. Then tenderfoot guests became more common. "As transportation became less precarious, the steady influx of visitors to the ranch became a matter of some moment," recalled Cleaveland of early twentieth-century New Mexico. "The invited, the uninvited, and the self-invited came in unbroken procession. I think every rancher of that era will bear me out when I say that the food bills averaged twice the normal amount required by the household, and many a little ranchman has given his guests credit for his bankruptcy." If visitors strained the budget, they also provided entertainment: "It was tenderfeet who were the raw material for most of our jokes, practical or otherwise." Thus I never feel bad imposing upon the Cowdens' hospitality—after all, I do ride for the brand as well as provide comic relief. Just watch the creaky cowpoke haul his ass up into the saddle on a reasonably tall horse or climb down after five or six hours riding the range. "Feel short, do you?" asks Herman, all smiles.

Along with grub line or tenderfoot guests, visitors from neighboring ranches came for occasional dances, more elaborate than the deal Sam arranged after branding a couple of years ago, when Leigh Ann brought out her guitar and Pete fried fish and testicles and Sam cut the two-step with Kathy and Abby and Hannah while the rest of the outfit sat on their heels. Old-time ranch dances had cowboy after cowboy lining up to squire the few females available, so women wound up the evening—dances often went on until dawn—utterly exhausted. "The dances started at sundown and continued until sunup," Cleaveland wrote. "There were at least four or five 'gents' to every 'lady.' These latter ranged in age from Grandmother to little pigtailed Susie; anything feminine would do as a partner. . . . Toward morning the sufferings of my sex would become acute. Dazed with fatigue, I have dragged myself through the last hour of many a dance, praying for the sun to rise and put an end to my misery." Back in 1853, John Sherburne—part of the Whipple expedition that crossed the Cowden Ranch before reaching Anton Chico, the first settlement after weeks of travel across the plains—recorded in his diary this account:

Started early this morning & reached Anton Chico. . . . In the evening they made a fandango for us. Of course we all attended & no one was sorry for doing so.

The senoritas were dressed finely, with dresses of silk & muslin, many with white dresses. Several I noticed with very pretty feet, enclosed in "Americano" gaiters of cloth & patent leather. These were probable brought from Santa Fe by their [illegible]. The dancing was very graceful, especially waltzing. They moved without a particle of exertion & seemed to take fresh life at every step. When the dance was over, the senoritas took one side of the room & the "hombres" the other. Very little or no conversation was carried on between the sexes, which was much more agreeable for me as I was an "ignoramus," as far as regards the Spanish language. . . . Several dogs were running round the room occasionally fighting, receiving a kick from some one. A large sheep was by no means the least amusing thing to be seen. He would chew tobacco in large quantities & butt everyone that looked hard at him.

If entertainment in the ranching Southwest seems rude then and now, an extension of a mostly humble home life, well, it goes with the territory, which lends an almost infinite perspective to human endeavor.

Tumbleweeds

Shipping 2001

A group of jolly cowboys discussing plans at ease,
Says one: "I'll tell you something, if you will listen please;
I am an old cowpuncher and here I'm dressed in rags,
And I used to be a tough one and take on great big jags.

But I have got a home, boys, a good one you all know;
Although I have not seen it since long, long ago.
I'm going back home, boys, once more to see them all;
Yes, I'm going to see my mother when the work's all done this fall.

D. J. O'Malley, "After the Roundup"

*O*ctober 18, 2001—Shipping season on the Cowden Ranch, and here's
your old cowpuncher again, ready to ride. O'Malley's cowboy dies
trying to prevent a stampede and never sees home when the work's all done
that fall; I feel like I *am* home. Sam and I once talked about cowboys going
nowhere, working long hours for short wages, "probably making eight
hundred bucks a month," he said. "And spend it all on their tack. Then they
hit a certain age and they've got a bunch of tack and they're beat up and
nobody needs them anymore. I've met quite a few over the years, and that's
what they do. Just go from one ranch to another, you know, try to see as
much country and ride as many horses as they can, while they're young.
But there's not many of those people around anymore."

Me, I'm trying to ride this particular country before I'm gone or it is.
My flight out yesterday was unsettling because of the September 11 attacks.
The trip went smoothly enough, although I got spot-checked in Hartford
and Cincinnati both, what with my computer and camera bag full of elec-
tronics. The woman who screened me in Cincinnati asked where I was

headed, and I explained my book, the ranch, cattle work. "Oooh," she crooned. "Why don't you send me back a cowboy?" The mystique survives.

I'm in the bunkhouse, which is still a mess from Geronimo's truck assault last May. A carpenter—former cowboy named Spike—has been rebuilding the wall and putting in new windows, so the rubble is gone, but all the bedroom furniture is jammed everywhere else. Doesn't bother me, and I'm the only cowboy here this fall; no one else is coming. Herman just returned from the Pecos Wilderness, deep in the Sangre de Cristo Mountains, where he got his deer. Besides Sam, Herman, and me, over at Alamito we have Jesús—Geronimo's older brother—riding for the brand; the four of us will handle things, along with neighbors on days we work the herds. Next Wednesday, Sam ships two loads of calves and we'll be getting everything ready. Sam is sorting his pairs—keeping the heifers, shipping the steers—and also sorting by breed and size. That's the task over the next few days, gather and sort—tomorrow we'll throw some cattle together.

It's good to be here in the fall. I left Santa Fe today in late afternoon, cottonwoods glowing golden along the river and around ranch houses on the way, the fading light casting long blue shadows, the skies doing their stunning sunset thing. I arrived in Santa Rosa at sundown, as if returning from a long, difficult journey. At the rodeo arena, I found the Cowden and other 4-H kids riding happily, so apparently far from all the tumult in the world. It's easy to believe the West remains a more innocent realm than the rest of America; perhaps that's not inaccurate. And the pressure to preserve innocence seems ever more immense.

October 19, 2001—Friday, first day in the saddle. Sam's trophy saddle, in fact—"Santa Rosa Roping Club, 1999 Champion Heeler" is tooled into the leather, along with basket-weave stamping on the skirts and fenders. More important than decoration to a working cowboy is "good wood," a saddle that won't wear out horse or rider. Sam's trophy is a roping saddle: fifteen-and-a-half-inch seat on a Bowman tree, double-rigged with a low horn and cantle. Like many trophy saddles, says Sam, it isn't expensive or even particularly well made for ranch work, though it's comfortable. Big Guy's and Rooster's rugged old saddles sit in the barn, and Sam says they're rough rides. Cowboys can debate tack almost endlessly, many of their preferences and prejudices based on utility, some on style. In general, roping saddles are designed to shift the rider forward and keep him there; such rigid positioning may assist the act of roping but doesn't make long rides easy. On a ranch saddle, "the seat is different, the rigging is a little different, where your cinches are placed on the tree," Sam says. "It lets you have more freedom with your legs. A roping saddle is rigged so you don't have much movement with your legs, to force you to the front of the saddle and rope. Ranch saddles are designed more to sit down in, for comfort, to cruise, to be able to stay in it all day. They're deeper, probably better for riding colts or something, if one did buck." However tough and uncomplaining cowboys might appear, they can still feel a bad saddle or a rough horse.

Sam had me ride Hollywood, an honor, although last spring he came up lame, didn't get ridden, and put on a few pounds. When I saw him yesterday, I teased Sam: "Damn, *primo*, I open this book with Hollywood—top cow horse, hero of the West—and here he is sporting a beer belly. Good thing I took his picture two years ago." Truth is, horses can run afoul of any number of problems—snakebite, barbed wire, prairie dog hole, old age, whatever—and you're lucky if they all remain healthy. This afternoon we drove into Santa Rosa to pick up a mare Sam wanted to try for Hannah, who is not yet a perfectly confident rider and needs a gentle horse. When we got back, Sam wanted Kathy's opinion, and she rode the mare. "She's fine," Kathy reported. So Sam might buy her; his only hesitation is that she's been getting special feed and daily care on her eyes for some problem. She might require more attention than horses receive here, or she might adapt and become a satisfactory ranch horse. With her, I count eleven horses in the remuda: Hollywood, Pajamas, Louie, Crow (a white horse, but jet black as a colt), Streak (Herman's horse, who was cut so badly running through a barbed-wire fence), Dunnie, Red, and I'll have to get the other names later. I'll ask Sam to write down some description of them, or get the whole family on tape, talking about the horses, and then get Sam's private assessment.

So I was on Hollywood when Herman, Jesús, and I left headquarters this morning. Sam got held up, speaking to someone about the new horse. "Isn't that just like el jefe," I told Herman. "We've got to ride hard while he makes phone calls." "Oh, I know it," Herman said, "I'm used to his tricks." We began throwing the cattle—scattered throughout the Shipping pasture—to the east. Herman took the outside, I followed, then Jesús, and we moved the cattle away from the west fence, picked up those in the southwest corner and along the south fence, and pushed them toward the windmill near the east fence. "We're going to have a shitload of cattle," said Herman when we got together; cows and calves were moving from all over toward the rendezvous point. Sam joined us, pushing more cattle from the north fence, and we had the bunch together, 260 pairs strong.

We moved the herd to the corner for the sort, more complicated than usual because you must bring out a pair, not one animal. "First you have to identify the pair," said Sam. "But you've got some clues. You know any Hereford calf will have a Hereford mama, and any black calf, an Angus mama. The black whiteface calves are more difficult. We'll let them mother up, but some of the cows have already started to kick off their calves." Sam started cutting, and the herd held easily, which isn't always the case. "Same bunch of cattle we had the other day in Conchas, but that day they were running all over the place. Our horses were worn out by noon. What causes that?" wondered Sam. I asked if it might be weather, and he allowed how it was cooler that day and maybe made the cattle flighty. Today was clear, still, and mild—at least seventy degrees—so perhaps that kept the cattle calm.

For a premium on the price, Sam had agreed to provide the cattle buyer with a select load of Angus or Angus-cross steers, which would eventually sell as Certified Angus Beef. So our first sort was to take out all Hereford calves; after that we'd sort out heifers. Sam asked me to cut out a few. Hollywood and I moved into the herd, which was holding well—not too tightly bunched, standing idly as if in deep thought or no thought at all. I wanted a pair close to the outside, rather than deep in the bunch. I spotted a red calf beside its mother and put Hollywood on them; he began moving them out through the other cattle, slowly and without apparent effort. I could have laid down the reins and just let Hollywood do the trick. That's what happens in cutting-horse competitions—you are just along for the ride. In this case, the process becomes complicated only if the calf wanders away from its mother; then you have to keep both calf and cow identified, bring them back together, and move them out. At one point Sam gave me welcome advice: "Take out the cow, primo—the calf will follow." Duh.

It's not often Sam gives instructions, unless requested. That violates code, which calls for new hands to pay attention and learn the ropes by observation. Rooster was famously tight-lipped. At one point when Walter Kellogg was on the ranch trying to lend a hand, he protested, "Rooster, you're keeping secrets from me!" In *No Life for a Lady*, Agnes Morley Cleaveland wrote, "Most of our real work was done in silence. Tenderfeet always complained, with justification: 'Nobody tells me anything. How am I to know what to do?' Kibitzing was our greatest social sin. A stranger could be with a cow outfit for days without knowing who was boss. Nobody shouted orders around a cow outfit. The boss merely picked up his bridle and started for the corral. Sometime during the process of saddling he managed to convey to the other riders that the drive would take in such and such territory and that so and so could work the *rincon* or west pass or piñon flat. And the men rode off in silence, each knowing what to do and needing all his resources to do it."

But I was glad to have Sam's help, and that of Hollywood, who resisted my attempt to press that first pair forward more quickly; he knew not to rush. And once we had the pair ready to move away from the herd, Hollywood held back, daring the calf or cow to make a break, as Sam said after: "You see what he was doing? He was baiting them, giving them room so they'd try him. I've had him do that lots." Sam also said it's harder on a good horse like Hollywood to cut pairs, because his instinct and training is to get onto one particular animal and take it out. With pairs, he's got to check himself. It was great fun for me. Like Hollywood, I developed tunnel vision to keep the pair in focus among all the other cows and calves; the whole herd and other riders virtually disappeared as we worked. Even though he was out of shape, Hollywood had all his experience intact and could do the job easily—sorting requires judgment more than muscle.

After I cut out eight or ten pairs, Sam and Herman cut out additional pairs, trying to leave behind as uniform a calf crop as possible—nothing red, nothing small. That way these steers would move smoothly together through the chain that eventually puts a choice steak on your plate. For a good while I "held cuts," keeping them from returning to the herd or drifting off too far. If the cuts aren't reasonably close, cattle being sorted are reluctant to move away from the herd. You want to hold them close enough to attract new cuts, but not so close they try to return. "I *hated* holding cuts as a kid," said Kathy later. "So did I," said Sam. "Sure dull to a kid. Now I think it's not so bad to take a break."

When the sorts were done, we moved the heifer pairs up the fence line to the spring in South Sabino, then rode home, back by 12:30 or so. By that time I was stiff, which is the real nature of being saddle sore. I also got rubbed from knee to ankle, and my shoulders and arms were surprisingly sore. Fun though. Hollywood was easy riding, with a smooth slow lope when we were pushing cattle, a more comfortable gait than any trot. He got a little tired but always came up when I asked for it. Hope I get to ride him tomorrow, although Sam hasn't wanted to use him two days running because he's out of shape. Kathy and Herman tease Sam about spoiling Hollywood, still his sweetheart.

October 20, 2001—Up on the mesa outside Sam's new trap, a small pasture used to sort or hold cattle for short periods. I'm waiting in the pickup while Sam, Herman, Jesús, Abby, Hannah, and Guyito sort a bunch of forty-three pairs we brought up from Alamito. I rode with Herman and Jesús on the drive, Sam and the kids met us, and I turned my horse over to Guy. Sam wants the kids to learn cattle work and this is a perfect opportunity—instructive but not so demanding they can't handle it. In fact, it's the boss who is holding cuts. Herman had time to identify the pairs as we moved the cattle, so he's sorting; Jesús and the kids are on the perimeter, helping to release and seal cuts from the herd. Sam was careful to explain to me he was short on mounts and wanted to let the kids ride. I'm not beefing—I ride or, in this case, don't ride for the brand.

How did Sam get short on horses? Hollywood isn't 100 percent and Sam sold Macho and Pappas, who were on the ranch in May. "I'll get even with *mi cuñado* one of these days," says Sam of the broncs that brother-in-law Souli sent him. *¡Vaya con Dios, Macho! ¡Vaya con Dios, Pappas!* And Louie is so crippled he couldn't possibly be ridden, can't even limp in to feed. He stays up on the mesa in the horse pasture, hardly moving. The problem's in his right foreleg, which doesn't look swollen, though he has a slight cut. Something is wrong, possibly resulting from a fall when he was unloading from a wet, muddy trailer in September. He's always had trouble backing from a trailer and he's too long and tall to turn around—bad procedure in any case—and he fell, then came up lame two days later. No one knows the exact injury, and Louie's not talking. *¿Qué pasa, Louie?*

I rode Pajamas, who was fine—Sam uses him for ropings these days—but not nearly so smooth as Hollywood. You get to the point where a horse becomes an extension of you, as you get cattle to do your bidding. Lean a little to the left, lean to the right, and your horse turns that way, hazing a cow back into the herd. I recall Souli training his paint colt, teaching him to rein and react to cattle, Souli not *thinking* about the horse but *willing* him to react with rider. "Hell on horses," and perhaps the better the rider, the more dominating—not so much by force as by will and understanding, since no good cowboy likes to see a horse mistreated. I do admire the way horses look and move, their nobility, their abilities. They are more "masculine" than cattle, which are actually females mostly, just as most cow horses are males. But it's more than simple biology—the strength and utility of the whole species tilt horses toward the masculine. In Spanish, it's *el caballo* and *la vaca*; the horse is masculine, and the cow feminine.

So they'll finish this little sort without me. Like yesterday, heifer pairs are going into South Sabino, Hereford pairs over to Creek pasture, and Angus steer pairs down to Shipping pasture. This new trap is well located, almost dead center of the ranch, letting Sam gather or scatter as necessary.

He told me there were few fences on the ranch at first, and no pens at all up on the mesa. North and South Sabino were one pasture, over twelve thousand acres. The Cowdens had to drive cattle from up near Little Conchas windmill all the way to pens at Sabino Springs. That's probably twenty miles of riding before you even hit the ground to work. I checked my map program and figured I rode over ten miles on Friday, and half that this morning. Most of today went at a walk, for which I was thankful. Ease the new/old puncher into his routine; let his saddle sores, his tormented bones and joints, rest a little.

Now it's after 9 P.M. and I'm in the bunkhouse, happily settled into the sack, making notes about the value of hard physical labor combined with this climate and geography. I go to bed very tired and sleep well, even after a good siesta in the afternoon. Of course I'm rising before six, riding miles and miles in the fresh air and sunshine, having a full meal at noontime . . . why wouldn't I fall to sleep, afternoon or night? Add in the change in altitude, ascending a mile into the atmosphere, air thinner and light clearer; add in sun and wind; add in a due vigilance about my surroundings—and what results is a proper exhaustion. Or not exhaustion, nor weariness, but simply a tired person, physically and mentally and imaginatively. Tonight once again I expect a long, deep sleep that leaves me refreshed, even if my body is sore and stiff from unaccustomed labor. Riding in the pickup with Herman today after sorting, I admired his fitness at forty-seven, his physical soundness and positive outlook. From character or surroundings, nature or nurture? There have long been claims for the healthy life on the prairie, and I suppose in many ways they are true, though not to the degree claimed by promoters trying to make a buck on a book or a rough patch of ranchland.

Tonight after supper we had a campfire in the pit Sam built near the front porch. Campfires are one form of entertainment here; Sam's friend Pete Marez, the animal control agent, said that his survival training included guidance on being lost in the wilderness. Your first objective is to find a road, then to build a fire and stay put. The reason you find a road is obvious, and the fire will keep you from freezing but also provides some companionship for those lost in the wilderness. Just having a fire keeps the lost soul together. I buy that, and why not? My sister Suzanne once proposed that, to understand women, men should tend fires; to understand men, women should ride horses.

So we sat around the fire, getting in touch with our feminine sides and swapping stories. Sam said last fall they built a fire here and cooked a brisket and ribs down in the coals, baking them slowly overnight . . . so they thought.

Turns out their dog Pepper, despite hot coals, dug and dragged out the meat and ate the whole feast. Rib bones were scattered everywhere in the morning. Pepper's an equal opportunity thief, because when hunters were staying in the bunkhouse during antelope season, she nosed open their cooler and stole a brisket and rib-eye steaks. Sam guesses it's brisket that excites her, since antelope meat was readily available, none of it buried under ice inside an Igloo. Sam was forgiving enough then and amused now; after all, Pepper *is* a dog, and meat *is* good.

What to think of ranching in the twenty-first century, with vegetarians and environmentalists haunting Sam's dreams? America still is, as a new book chronicles, Fast Food Nation. MacDonald's and Burger King are still sizzling hamburgers, although die-hard ranchers complain that far too many chicken sandwiches are on menus now. Cattle ranching is under pressure, but I don't believe critically imperiled—people have to eat. Yet like anything that requires much space in a world shrinking under human sprawl, ranching and other forms of agriculture will see costs rise. What does a steak now cost in Japan? What did land cost immediately after the open range closed—fifty cents or a dollar an acre? What did this ranch cost fifty years ago—ten dollars an acre? What would it cost today? And what about the relative prices of cattle and equipment over time? Sam likes to calculate the cost of a pickup in terms of calves; in 1950 you sold fifteen steers to buy your Ford; today it takes fifty. Of course, the modern pickup is far more advanced than its predecessor and cattle have not evolved as quickly or dramatically as technology, so you can't expect cattle prices to keep pace with equipment costs. But Sam has a point—ranchers certainly aren't gaining economic ground.

He goes on to say, "The manager of Singleton's division here told me they can't cover costs. Own the land, own the cattle, and still can't make a profit on the operation. Something's wrong." What may be wrong is the position of cow-calf ranchers in the market stream. Sam owns a cow that has a calf, which he sells at weaning; that steer goes to the Midwest for additional growth on pasture, then to a feedlot. After that comes packing plant, wholesaler, retailer, consumer—six degrees of separation in most cases. Sam gets less than one dollar a pound, and a Certified Angus sirloin sells for thirty dollars at Ben Benson's or Gallagher's Steak House in New York City. The two poles of the marketplace are far, far apart, and ranch economics continue to change. Nineteenth-century cattle kings made fortunes from free grazing, cheap cattle, and cheap labor; none of those elements now apply. Land, capital, and labor have all grown expensive, particularly land and cattle. Labor costs figure prominently in most realms of production, but ranch hands, unskilled or experienced, aren't highly paid—one of the few economies cattlemen can expect. Yet labor is scarce, and most practicing ranchers—like Sam rather than Singleton—must depend on themselves. "Bill Mitchell manages the Moeses Ranch west of Santa Rosa," says Sam. "Does it all himself. Every bit of it. Most everybody is in the same boat. Even Singleton's operation is shorthanded; they can't keep anybody around. Division manager Alex Carone is on one ranch; three buckaroos hold down the Conchas; one old guy is on the Bar Y, another on the Latigo. Something like a dozen men for half a million acres. Not good."

Sam has a dream, long dreamed and sometimes practiced by his ranching forefathers, of running a ranch with just family resources. "If I could get this where me and Herman and the kids could handle it, that would make a world of difference." The recent parade of temporary hands—I've seen four in the past couple of years—is frustrating. Jesús is twenty-one and certainly a better hand than Geronimo, but he isn't a solution for the long term. "I made the mistake of going into the house at Alamito," Sam said. "The place was a complete wreck. Heater on full blast, windows wide open, beer cans and cigarette butts and trash everywhere. Mexicans used to keep things in good order, but no more. And you can't get them to stay. Work's too hard and long; pay's too low; it's too far out here. Heck, primo, I have to import cowboys from Massachusetts!"

October 21, 2001—Sunday. This morning the Cowdens drove into Santa Rosa for church—"Pray for me!" I implored—while I did research here on the Web. I wanted to see what a surefire buckaroo, or any self-indulgent cowboy, might spend to outfit himself and his best friend in the whole wide world. "I see by your outfit that you are a cowboy," goes "The Cowboy's Lament (Streets of Laredo)." I checked out the Big Bend Saddlery, CowboySuperStore, and other websites to burn my hard-earned virtual wages. Here are the results, from hat to horseshoes:

Hat: Both a straw Resistol "George Strait San Antone" ($70) for summer and an 8x felt Stetson ($250) for winter.

Bandanna: A classic red cotton one costs a buck, but Souli—who is tough and stylish both—tried to convince me last May that to protect his neck summer and winter in more comfort, a cowboy actually prefers a silk scarf, often called "a wild rag" ($24). You bet.

Shirt: Long sleeves to shield your arms from damaging sun and dust, with pearl snaps, $30 from Panhandle Slim.

Gloves: Cowhide, of course. $10.

Belt and buckle: Skip the turquoise and silver Navajo concho belt ($1,000) and get a simple leather model tooled with a flower design ($125). Use your own All-Round Champion Cowboy buckle or buy a handmade number with gold steer head and brand, gold rope edge, and antique finish ($325). Good enough for me, and affordable compared to one really elaborate model priced at $2,640.

Jeans: No contest. A Levi's bumper sticker reads "Wrangler Butts Drive Me Nuts," but Levi's are California gold-miner material. Go Wranglers, $23. And don't tell, but Sam claims some cowpunchers wear silk long johns or pantyhose(!) under their jeans to prevent chafing ($5 for L'eggs Sheer Energy).

Chaps: Pronounce that "shaps," and get full-length batwings ($295–$425) for brushy country and short chinks ($275) for hot, open plains.

Boots: Lots of choices and debate here, but the soles of manly footwear must be leather, so you don't hang a foot in your stirrup and get dragged to death by your horse. (Some cowboys used to pack six-shooters in case that happened—to shoot their horses.) Because they're comfortable and cheap, I'm wearing plain black lace-up Justin cowhide ropers ($100) that I don't mind trashing in the pens. For show, I might sport a pair of Lucchese Cognac Full-Quill Ostrich dress boots ($500), "handmade since 1883" in San Antonio.

Spurs and straps: More buckaroo jewelry. Try a pair of tooled-leather straps ($125) and either simple factory spurs ($45) or nice ones of browned steel with hand-engraved silver overlays, available with brands or initials in place of the outside silver mounting: 1-inch band, 2-inch shank, 12-point/1¼-inch rowel ($265).

Saddle: From Big Bend Saddlery in Alpine, Texas, since 1903, any-where from $2,000 to $4,000, depending on design and trimmings. How about a Will James model, basket-stamped with flower-carved corners, semisquare skirts, and full double rigging with stainless steel Ds? With a small Cheyenne roll, 3½-inch horn, 2½-inch silver horn cap, and braided rings on the back strings: $3,315.

Stirrups: Choose basket-stamped half-covered stirrups with your brand or initials ($80) or heavy cast-brass oxbow stirrups ($75). Or maybe tapaderos, a.k.a. taps, which are enclosed in front and protect the foot thoroughly—leather Carlos Border Monkey-Nose Tapaderos run $175.

Blanket: A good, "using" wool blanket next to your horse's back ($50) and a state-of-the-art Ultra Pad ($150) that will further protect Trigger and keep your saddle from slipping.

Bridle: A set might run $425, or you could mix and match fancy woven horsehair headstall ($225), Les Vogt browned-steel and silver Texas Star snaffle bit ($289), and harness leather reins ($40)—split rather than one-piece so, should you drop them, your horse will step on them and stop, instead of run away with or without you. (Some horses, however, learn to run with their heads turned sideways so they can avoid dropped reins. The old notion that a cow pony will stay "ground tied" is best avoided unless you don't mind a long walk home.)

Breast collar: To keep your saddle from shifting back, a three-ring model, basket-stamped, with stainless hardware—going for $65 to $140.

Horseshoes: Your blacksmith, or, more precisely, farrier, will pedicure your horse for $50.

Accessories: Your Cactus or Rattler nylon or poly rope ($25) and leather hobbles ($40) are really necessities. But tooled-leather saddlebags ($165–$235) are not, nor is a cowboy bedroll tarp ($144), so you can snuggle up in your soogans by the campfire like an old-time waddy. If you ride somewhere it actually rains, tie on your slicker ($30).

Total damages: $7,087. I skipped the saddlebags and bedroll. Price does not include horse ($2,500–$10,000), Ford F350 Powerstroke diesel dual-wheel pickup ($30,000), and four-horse Sundowner gooseneck trailer with stud divider, mangers, walk-through door to horse area, awning, stainless nose, boot box, and closet ($24,000, used only slightly). So it looks like I can ride the range, punchy as can be, for under $70,000. On a cowboy wage of $800 a month—assuming they provide room and board in the bunkhouse—I can pay off everything in just seven or eight years!

Unlike my virtual buckaroo, cowboys who rode for the JALs or other big outfits were supplied with horses and rarely indulged in luxuries—just as today on the Cowden Ranch, Sam provides Herman and any other full-time ranch hand with necessary equipment. But not the neighbors who come to pitch in. Those saddle bums, like me, get a good meal and a lifetime of memories.

Late this afternoon after we fed the horses and caught up our mounts for tomorrow morning, Sam and I drove into Bogie and up Esteros Creek. It wasn't running, certainly, but the winding arroyo, cut over the centuries, was clear enough and spoke to its long life. Every place on the ranch where live water once flowed is marked by human habitation. Rock-house ruins stand at Little Conchas, Sabino Springs, and other canyons on the edge of the mesa, and along the headwaters of Esteros. We stopped at the first ruins visible from the road, and Sam pointed out the typical, small remains of an outbuilding: "Someone told me maybe a storehouse. It wasn't an out-house—no hole." The house had a main room and two or three other attached rooms. I marvel at the lives centered there, undoubtedly sheep-herders bringing flocks to graze from spring through fall. At most sites are stone corrals—one large corral farther "upstream" measured about a hundred feet square—with small attached lambing pens, each about the size of a baby crib, where ewes were placed to give birth in safety. We visited a handful of different locations, many with an old cottonwood or willow to indicate there was once more abundant, regular water; elsewhere on the dry, rocky slopes, only junipers and piñons survive.

We explored Esteros until its sources played out—rock seams in a hillside where springs occasionally surface—then Sam drove up on the mesa to a point above the headwaters. The sun was dropping and gold light suffused the whole landscape. To the east, Esteros wound its way off the Cowden Ranch before turning south toward the Pecos; in the distance the Llano Estacado was visible as a blue rampart, and other detached mesas stood in the last flashes of light. To the west lay the wide open valley where Alamito camp sat in isolation. Well beyond flowed the Gallinas River, and beyond that rose the Sangre de Cristo range, where the sun set. Clear and beautiful, all of it. The low ridge between Bogie and the western valley was *el paso* that I believe so many used to travel between plains and mountains—Pueblo, Apache, and Comanche Indians; Coronado; Hispanic ciboleros, pastores, and comancheros; Gregg, Marcy, Whipple—yet nothing was evident of all those journeys and lives. Only Sam and I, watching the landscape change at sundown and dusk, light and shadow clarifying every cedar, each boulder and ruined stone house. For his whole life Sam has lived on this ranch, but such visions still produce in him the same sense of awe I felt.

Now I've got the bunkhouse to myself, and I'll turn in early. Big week ahead. It looks like I'll stay until next Saturday. After Sam ships his calves on Wednesday, we'll start preg-checking—or "palpating," the medical term for examining by touch—all of the cows. With some additional help, Herman performs that task, which he learned at Sam's encouragement. Sam wants Herman to feel he's making progress—thus palpating, an artificial insemi-nation course, his own cows—and not simply riding himself into the ground. Herman seems content, however, firmly committed to the ranch. If there was ever a man who rides for the brand, it's Herman Martinez.

We'll sort again tomorrow. Then Sam and I will go help Joe Tom White work his cattle on Tuesday; the same buyer who's taking Sam's steers is buying Joe Tom's. On Wednesday we'll ship the steers and test that bunch of cows. On Thursday we'll get up the pairs in South Sabino and sort off thirty or forty calves Sam wants to keep as replacement heifers, then preg-check those cows as well. It will take the rest of the week to get the cows checked, calves vaccinated, and so forth. Sam asked how long I could stay, because with my help he could get all the cows worked. A full week ahead, lots of riding, and the promise I'll work myself into top shape. Said Sam, "You'll be a lean, mean cowboy machine."

October 22, 2001—Monday. This morning we gathered Shipping Pasture and sorted one more time, moving off the last of the Hereford calves and heifer calves. Sam is left with about 195 cows and has 190 calves to ship on Wednesday. There will be little sorting that day, which stresses the cattle less and helps avoid working weight off his calves. I also think he likes sorting in the pasture, on horseback, the old way.

I rode Hollywood this morning, and we all did fine. The cattle gathered easily and held well along the east fence; they didn't want to move to the corner where they've been sorted so often, but no real trouble. We took out ten cuts and seven heifer-calf pairs. A steady northwest breeze blew and Sam cut into it, as Charles Goodnight suggested was best—that way cattle can scent what's ahead. Seemed to make the job easy; we were done by 10:30. Then Sam and I drove up to South Sabino to make a count and feed a little cake; Sam wants to accustom the heifer calves to feeding since he'll start just after weaning. It's important for replacements to get adequate nutrition to grow well this next year, so they'll breed readily and keep those future generations coming.

While we were in South Sabino, Sam told me the story of one October a few years back, when Souli and Jake were both here to help ship. Also visiting were some cattle feeders from Kansas, who often bought Sam's calves and wanted to see his operation. Sam and the Kansas men were holding herd near Sabino "lake"—at the time nothing but a large, mostly dry depression in the pasture—while Souli and Jake were bringing in the last of the cattle. "That's what I thought anyway," Sam said. "Turns out those crazy Shanklins ran across the antelope that always hang around near the playa, and damned if they didn't put their Thoroughbreds on them. You know how antelope will run on you. Souli and Jake got about a dozen going and here they come over the rise, two cowboys running a herd of antelope full speed, right past me and those cattle feeders. You should have seen those Kansas boys' eyes. Souli and Jake never stopped, raced the antelope down one side of the playa and out the other in a cloud of dust. You knew what the Kansas feeders thought: crazy Wild West cowboys, all of us!"

A couple years later, Sam went with Joe Tom White to the Gardiner Angus Ranch in Kansas—a renowned purebred herd—for its bull sale. "Well, folks had heard from the cattle buyers all about running those antelope," Sam told me, "and we were instant celebrities. Those Kansas farmers looked at us like we were going to ride into the saloon and shoot up the place." But Sam wasn't looking to rattle the ranch or even to buy cattle. "First thing they ran through the sale was a cow, a *dry* cow," he said, "that brought eighteen thousand dollars. She was a donor cow, one they used to AI [artificially inseminate], then strip the embryos and plant them in other cows. She was quality stock and humongous, probably eighteen hundred pounds—she'd never raised a calf. But still, eighteen thousand dollars for a dry cow! I said, 'Joe Tom, when an old dry brings eighteen thousand dollars, we ain't got no business being here. Let's go!' We had a big time."

Despite high prices, improved cattle—primarily bulls—from midwestern breeders routinely arrive in this area for sale. The Oklahoma cattle buyer taking Sam's calves, Lusty Reyher, has sent out truckloads of purebred Angus bulls, all developed through artificial insemination by performance-tested sires and dams, to Sam and other area ranchers. "Lusty buys a lot of cattle around here, several thousand each year. Lots of them from small-time Spanish raisers, and they are good calves, in part because of these quality bulls," said Sam. The most efficient, quickest way to improve a cattle herd is through the sire, since one bull breeds twenty or more cows each season. A small, part-time rancher can buy one bull and influence his entire calf crop. "Just last week a bunch of people brought their calves into Santa Rosa," Sam said. "They had seven hundred head in the arena there. Lusty told me the calves were fifty pounds heavier than last year's bunch—because of good bulls." Sam buys AI-bred Angus bulls from Lusty and other suppliers; some of the bulls are developed as far away as the University of Michigan. The science of breeding has taken enormous strides, intensifying a trend that began long ago when early cattlemen imported British breeds to improve Texas Longhorns and other native stock. Over the years, breeds have improved internally, and new breeds have been developed to meet range conditions, cattle buyer and consumer preferences, and the needs of producers, who ideally want a live calf every year that weighs and grades well. Both producing and marketing cattle these days can be dizzying.

On a related and sober note, Sam said that last week he shipped a mixed load of open heifers, late cows, and other culls. "I sent them to the auction barn at Dalhart, Texas, and they didn't do very good. There was a big scare that day. Turns out some Arab guys in a beat-up station wagon were driving around the Oklahoma Panhandle and Kansas, that area, going in and out of feedlots. Word got around on the ag network—all these feedlots subscribe to a service, me too—and everybody got worried. They could have been scattering anthrax or mad cow disease or something. I don't know how many feedlots these guys visited. Bulletins went out all over, and they were trying to locate this old station wagon, catch these guys. Never did. I haven't heard anything new on it. That was the day I sent my cows to Dalhart, and the market went to hell. Isn't that crazy? Lusty told me calves that day dropped to seventy cents from over a dollar [per pound]. It hurt me, for sure. I can't quite say how bad. Crazy."

Market jitters are understandable, of course, following the September 11 attacks and the outbreak of anthrax in Florida. Anthrax is an ancient disease of livestock, endemic in some areas since spores can live in the soil for decades. The form of the disease that cattlemen have always faced is, however, less dangerous than versions posed by bioterrorism. Those threaten human life and the whole livestock industry. The towers that fell in New York City created shudders worldwide. Despite the apparent isolation and safety, you cannot hide on a ranch in New Mexico.

October 23, 2001—Big day. Woke at 3:45, started the coffee, dressed in the dark, went outside, and immediately saw a shooting star. Countless others still spangled the sky—perfectly clear, not so windy, cold. Sam came over about 4:00. We talked about the swing in temperature this time of year: from thirty degrees at night to seventy in the afternoon. We start off each morning wearing gloves, down vest, and jacket, then strip down as the day warms up. But the weather could turn at any time, cold for the entire day. I just hope we don't get a string of rainy days that trap me out here; one October Jake got stranded for a week.

We saddled Hollywood and Pajamas under the stars—two days in a row on Hollywood! It's good to get used to a horse, especially one that rides so nice. Doesn't walk particularly fast, trots about like other horses, but his slow lope in the pasture is tops, to say nothing of his sorting cattle. I'm legging up, as Sam says. Not so sore today, though we didn't ride much yesterday. You just get settled, riding more easily and comfortably day by day. In touch with your equine side.

We took off by 4:30 for Joe Tom's; Joe Tom also sold his calves to Lusty. "You will like Lusty," said Sam. "He's a one-eyed bandit from Comanche,

Oklahoma. Patch over his eye. A good guy." Lusty? Eye patch? Were it not my true, straight-shooting cousin, I might suspect a setup. Practical jokes have been a staple of cowboy humor forever, especially when a greenhorn—from the term for a young animal with new horns—is available.

Sam and I arrived at Joe Tom's ranch east of Santa Rosa just before the appointed hour of 5:30. Lots of pickups and horse trailers were already there; we walked into a house full of cowboys and the smell of bacon and eggs, sausage, and biscuits that Karla White had laid out. By day's end I met or reacquainted myself with most of the outfit.

Ethan Fuchs was there with two of his five sons. Ethan practiced law in Phoenix, made a killing in real estate dealings—trailer parks and trailers, soul-deadening profitable work—then cashed out and bought a small ranch adjoining Joe Tom's. When Ethan and his wife had difficulty conceiving, they adopted a boy, then quickly had four sons of their own—their place is wild with little cowboys. Ethan told me, I suppose because I live in Massachusetts, that on his mother's side he descends from Cotton Mather, the Puritan minister. Religion also runs in Ethan's veins; he and his wife are devout members of the Assembly of God, which Sam tells me is the largest Protestant congregation in Santa Rosa.

Ernest Copeland—the preacher who often helps Sam brand—was on hand to pregnancy test. For a long time he preg-checked for the Bell Ranch, the storied land-grant ranch along the Canadian River about forty miles from Sam. Ernest bestowed the blessing before meals. What *is* this connection between piety and breeding cattle? With him was his daughter, Sandra, who last year quit her job, gathered her funds, and drove around Australia for a year. Sandra can ride and work on the ground with the best of them. Ernest is about sixty, serious, and left-handed. That's not insignificant, since most squeeze chutes for cattle are set up for right-handed cowhands. More on the intricacies of palpation later.

Not everyone was so pious. Steve Kimbrel is a longtime friend of Sam's, an Air Force brat—Cannon Air Force Base in Clovis, New Mexico, is a major field dating from World War II. Steve took up cowboy life and used to work for Joe Tom. He recently bought a little place near Tucumcari, where he raises alfalfa and does daywork as a cowboy. Back in their bachelor days, Sam says, Steve raised hell and chased women all the time. One morning when Sam was staying with Steve, he asked what was for breakfast. Steve tapped out some flour onto a plate, took his finger and divided it into three portions, and said, "Bacon. Eggs. Biscuits." "Steve was always broke—spent every penny at the bar," Sam says. "He was working for a rancher who was buying loads of calves, working them, and turning them out. Steve was living on calf fries. All he had to eat. The only food in his house was calf nuts, grease, and flour."

Bill Mitchell single-handedly runs a ranch for the Moeses family, who own lots of property in Santa Rosa and nearby. Sam tells me that once the— is it Moeseses?—forgot to record which customer in their feed store had ordered a new saddle. To solve their confusion, they simply billed all of their customers, figuring the actual purchaser would come forward. Trouble was, lots of people, thinking they ordered the goods, simply paid the invoice. Sharp storekeepers, trusting ranchers. Bill Mitchell is no merchant; he's all cowman. He pregnancy-tests also and won't take pay (I think Ernest gets one dollar a head) except for the reciprocal help that neighbors like Sam and Joe Tom provide. Bill's personable, didn't at all mind my teasing about his short-legged horse. He smiled, "I saw you trying to crawl up onto Holly-wood this morning. What did it take, two or three tries? I got tired of hunting a ditch for my horse to stand in while I mounted."

I also met local cowboy Fernando Madero and Tim Lamb, who runs a cattle operation for someone south of town and shoes horses on the side. Both Tim and his father, Irvin, previously worked on the Bar Y; Irvin was the manager and an important influence when Sam took over the Cowden Ranch. "We talked every day," says Sam. "Irvin really helped me out."

Not least was Joe Tom himself and his teenage son, Kasey. Another son is three or four years old; Joe Tom has had children in the 1970s, 1980s, and 1990s. As old Bill Cowden said, "Nothing to do but feed them."

That's all, maybe. We had fifteen riders gather a relatively small bunch of cattle, in part because Joe Tom's place is centrally located—it's not like driving twenty-six miles of rough road to the Cowden Ranch. Fall roundups often have a social element, and less work is required in the pens. On some ranches folks show up for a pleasure ride rather than actual cattle work, leaving after the gather and before any dirty work. We went out three different times to gather small pastures that surround Joe Tom's headquarters. Not much trouble, although some of Joe Tom's cattle have Brangus blood that makes them difficult. If you see a young cow get her head up and snort, then hold on—things may turn rowdy. As we neared headquarters, one fool cow got hot and broke from the herd; when a couple of riders went to haze her back, she lost it completely. The rest of us pushed the herd into the pens, and before long here comes the breakaway Brangus . . . with a metal Powder River gate hanging around her neck. She had gotten to a corner, started butting the gate to get through, slipped her head between the slats, and lifted it off its hinges. I wish I'd had my camera.

Other than that, the riding was a pleasure, not at all difficult. When we first started, Steve asked if any of us were riding a mare, because he was on a stud, which can occasionally lead to cowboy drama. "I wouldn't want him trying to mount your mare while you and me both were in the saddle," Steve said.

After the cattle were gathered, we started stripping calves from the cows, sorting steers from heifer calves, and weighing the steers. I met Lusty, who did have a black patch over his right eye. The left one was bright blue. He cut a handsome, dashing figure in his white Stetson and black patch, like a cowboy Captain Kidd. Beyond the pens under some big cottonwoods, two cattle trucks sat waiting for their loads. Joe Tom's ranch, like the town, is located in what is called the Santa Rosa Sink, a geologic formation of collapsed limestone notable for its many springs. Sam naturally envies the abundant water.

After the sort, while the crew weighed calves at headquarters, Steve, Sam, Ernest, Sandra, and I went to work a little bunch of cows at another set of pens. Joe Tom has one pasture on the other side of the interstate, and we drove over to preg-check. Ernest did the testing while Steve and Sam gave vaccinations and I punched cattle. Our only hitch came when one cow got down in the squeeze chute. Cattle have a habit of pulling back against the pressure holding their heads in place, and as this cow did, she went down, pinning her forelegs under the head gate. Steve tried to free her from the bind she was in, the cow slammed against the head gate, and, "Damn if she didn't just break a leg!" Steve lamented. The cow hobbled off toward a corner of the pens. Steve felt bad about the accident, which meant she would be sent to slaughter. "I hate for Joe Tom to lose a cow like that. And of course she was a big fine cow and bred too."

By the time we returned to Joe Tom's, they had the steers weighed—a fine bunch of calves—and we broke for a big lunch: chili, enchiladas, frijoles, biscuits, cake, pie. The talk was cattle prices, weather, and the Santa Rosa football team, which whipped Fort Sumner last Friday and had a chance to make state playoffs. Joe Tom's son plays on the team. After lunch, Sam and I were cut loose, along with some others; Ernest and Bill were staying to preg-check the big herd of cows. They had a full afternoon before them, with the prospect of another long day tomorrow, when Sam ships his own calves and tests his own cows. It's a busy season.

Sam dropped me at Jean's house while he went to talk with Lusty and run errands. I watered the horses and rested. When Sam returned, we went by the two feed stores in town, the first out on Historic Route 66 to the west, the other in the center of town. At one Sam got arm-length OB gloves for pregnancy testing; at the other we got needles, penicillin, plastic syringes, and ten 80-pound sacks of mixed feed for the weaned calves. And Sam bought me a nice wool saddle blanket to use the rest of the week, then take home to Massachusetts and "lay it beside your bed. You can get a little horse sweat mixed into that wool, so you'll have cowboy perfume all winter." A nice gesture, typical of Sam. Double-thick, blue with red and white highlights. One year after branding he presented me with a Moore Maker pocketknife, another memento of thanks. The tight son of a gun certainly isn't going to pay me!

The Feed Store is run by Don Sultemeier, and Sam tells a funny story on him. Don once donated a leather breastplate to a Santa Rosa benefit rodeo for a local man who electrocuted himself and had lost an arm and might lose a leg. Leigh Ann happened to win the trophy and noticed that the legend carved into the leather was misspelled: BENIFIT ROPPING. When she pointed out the error to Sultemeier, he declined to replace it with a correct version. Said he checked the spelling with his wife, who is a schoolteacher. Dubious. More likely a tight shopkeeper. So Leigh Ann still has a BENIFIT ROPPING trophy.

While in the store, we ran into Tom Payne, formerly Sam's neighbor to the west. Although Payne spent time telling Sam how much he had admired Rooster and what friends they had been, relations are cool. Rooster and Tom's father had an understanding that should the Payne ranch sell, the Cowdens would have first option on it. A long, narrow ranch that follows the Gallinas River for ten miles, it would have fit wonderfully with the Cowden Ranch, providing them reliable water on their west side. Patty and Souli had even talked of moving up from Texas when the Payne place became available. When that time came, however, despite promises to the Cowdens, Tom Payne sold to Singleton. Sam spoke courteously with him— Sam rarely speaks badly of or to any particular individual—but the code of the West requires a man's word be good. Enough said.

I'm tired but looking forward to tomorrow. Ship those steers—cash that check!

October 24, 2001—Shipping day, 5 A.M., still dark, handle of the Big Dipper pointing straight down in the northeast sky, wind blowing. We'll gather the cattle, strip the calves, weigh and ship them, pregnancy-test the cows. Bill Mitchell's coming to help Herman preg-check, Pete's coming, and we have Sam, Jesús, me—six hands. Yesterday we had fifteen for the same number of cattle.

Bunkhouse, 9 P.M. Tired puppy here, but happy. The day went well, cool without much wind. Sam had Jesús water down the alley yesterday to settle the dust, a good move. Everyone was on time: cowboys, brand inspector, Lusty, truckers. Lusty once told Sam that he figures a hundred people—those on the ranch and others back in various offices—are involved on shipping days. If the brand inspector doesn't show, for example, no cattle can ship. The rancher has just gathered and sorted for nothing, stressing the cattle, costing them weight on the scales, whenever that might come. On his ranch, Sam feels responsible for everything, and one hundred people is a huge crowd for him to rely on or coordinate. A simple cowpuncher like me never worries.

The big news is that Sam's calves, in Lusty's words, were "the best set of calves you ever delivered." Lusty added, "I'm not just blowing smoke up your ass, either." To be precise, 184 calves weighed 118,400 pounds, an average of 653 pounds each, and paid out $621 apiece, the most Sam has ever received per head for his calves. Lusty wrote him a check for over $93,000 for two truckloads. The trucks couldn't haul the whole bunch, so we had to cut back twenty-four calves. They are in the corrals right now, bawling for mama. I plan to drop off to sleep no matter.

Unlike yesterday at Joe Tom's, we had little trouble gathering the pasture, although they didn't want to drive toward the pens. We penned them, separated the cows and calves, started weighing. The aim is always to handle the cattle smoothly, with minimal stress. Sam's pens are sturdy and well laid out; I've sketched out the design, without showing all the gates—lots of well-placed gates make your pens work slick.

Herman and the other cowboys moved calves onto the scale—usually ten at a time—and my first assignment was to count them off, a double check. Inside the weigh house were Lusty and Sam, Lusty sliding old brass counterweights along the beam until each load balanced. "It's an old cowboy trick Rooster taught me," Sam said. "If you can get the buyer to weigh, it's customary for him to give you the benefit of the doubt, and round up. If the seller weighs, it's customary to round down." That crafty Cowden.

After weighing, I let the calves off the scale and made my count, while the New Mexico brand inspector checked their brands. I'm sure he wasn't expecting any trouble on this ranch—Sam trying to ship someone else's calves—but brands and brand inspections are required for all cattle that sell in the state. Inspectors are classified as law enforcement personnel, authorized and required to carry weapons. I asked Sam if that was why Leigh Ann, who was an assistant brand inspector for a long time, was

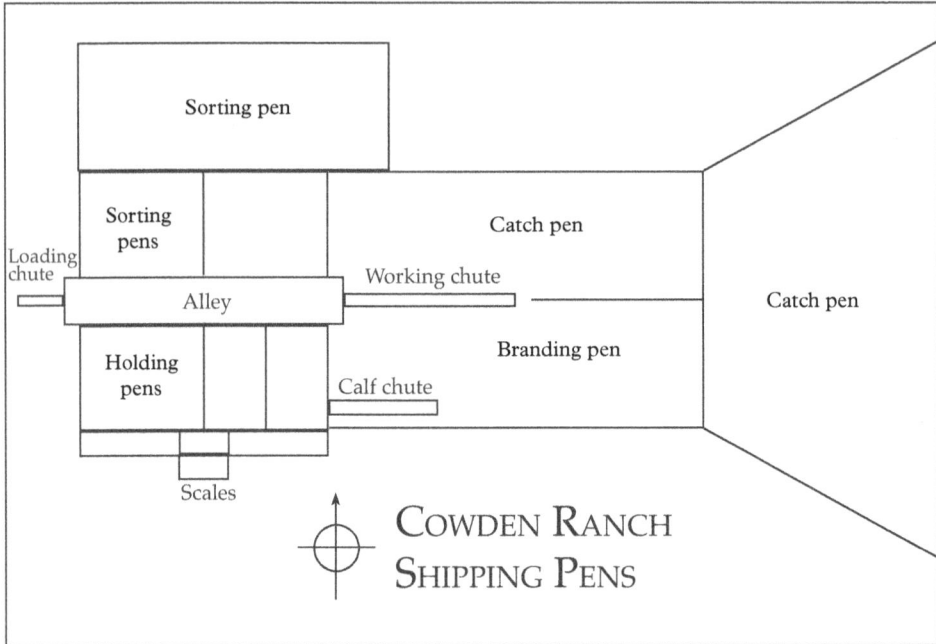

Sorting pen

Sorting pens

Loading chute

Alley

Working chute

Catch pen

Catch pen

Holding pens

Calf chute

Branding pen

Scales

COWDEN RANCH
SHIPPING PENS

passed over for promotion to regional brand inspector. Sam thought it was simple New Mexico politics; the job she was in line for went to a Hispanic man. Not because Leigh Ann is a woman, incapable of law enforcement—I sure as hell wouldn't want to get in her crosshairs—but because she's Anglo. The new inspector didn't look particularly qualified for police work; he weighed over three hundred pounds. In any case, all he had to do was check for the Lazy 6 on each calf's hip.

I told Sam what I'd discovered about the Cowden's place in the history of brand inspections, including the beginning of the Texas and Southwestern Cattle Raisers Association, officially founded in Graham, Texas, in 1877. Walter Cochran said that trail drives after the Civil War led to unauthorized gathering of cattle: "It was the intention of most of these men to pay for the cattle driven off and sold but good intentions do not go very far if they are never carried out, so that was the way most of these cattle were paid for—with good intentions." According to Cochran, the official organization followed an earlier meeting:

One hot day in July, 1876, a little bunch of cowmen met at Captain Dillahunty's store in Palo Pinto. They were talking about a certain man gathering a herd to drive to Kansas. Captain Dillahunty, who had had two or three drinks, mounted a goods box and said it was time this driving off of everyone's cattle was stopped. Colonel Kit Carter and Mark Lynn, two of the biggest cattlemen in Palo Pinto at that time, said they would commence then to see what could be done about stopping this driving off of everyone's cattle. Jerry Hart, Scott Warren, Ham Taylor, S. B. Strawn, Uncle Billie Cowden were in the bunch and seconded the motion. I may not be correct about this statement, but I will give it as my opinion that Captain Dillahunty was the real father of the Cattle Raisers Association from that little talk. The talk spread from one outfit to another, until 1877, a bunch of cowmen met at Graham and organized the present Cattle Raisers Association.

Other states eventually followed the Texas example and instituted brand inspections; the TSCRA today has livestock market and field inspectors ("Special Rangers") throughout Texas and Oklahoma to assist in prevention and recovery of stolen cattle. Texas registers brands by counties, but New Mexico brands are statewide, and Sam's single brand—Lazy 6—is unusual and highly valued here. Ranchers applying for new brands must frequently include three symbols, which makes branding more laborious. Leigh Ann has a book with all the New Mexico brands, and I want to see if the Cowdens' JAL is in there. Perhaps they didn't bother to register, since they were so far from Santa Fe and other state officials.

After a while, Sam asked me to take his place in the weigh house so I could see how the deal worked. It wasn't complicated. Little windows opened onto the scale itself and to each side; Bill Mitchell would drive calves onto the platform, close the gate, and call out the number of head. Lusty would adjust the weights until the scale balanced and ask me to confirm, "6,520 for ten?" We'd both write down the figures in our tally books. The calves were released, and the head count confirmed. After each fourth load or so, Lusty would call, "Balance," and we'd pause to adjust the scales while empty. My job was purely secretarial yet critical; on the open range, the tally man was a responsible and respected hand—Sam was honoring me. Big Daddy was tally man for the JALs, my grandmother used to weigh for Guy, and after her, Jean Cowden for Rooster. Kathy prefers to do something else.

Lusty and I had time to shoot the breeze. After periods in Colorado and Nebraska, he lives in Comanche County, Oklahoma, and ranges throughout the Southwest buying and selling cattle. As Sam said, Lusty seems like a good guy, though cattlemen traditionally suspect cattle buyers, who are often more informed about markets than producers and better able to make deals in their own favor. Though modern communications have bridged that gap considerably, buyers remain the butt of jokes: "Two mountain lions were trotting down the trail. Every time they came to a pile of elk droppings, the first lion would stop and eat a mouthful. Finally, in disgust, the second lion asked the first, 'Why do you keep doing that?' The first lion replied, 'Well, I ate a cattle buyer yesterday and I can't get the taste out of my mouth.'" Lusty liked that one.

I found out more about the deal Sam and Lusty struck back in July. In times past, rancher and buyer estimated what calves would weigh at shipping and agreed on a price per pound. Normally, lighter calves bring more per pound, so ranchers would estimate light weights, negotiate for a higher price per pound, and hope for heavier calves. Buyers would estimate heavy weights and negotiate for a reduced price per pound. The actual weight was a revelation that could make or break profits for one or the other. Nowadays all contracts are based on a "slide," which reduces the price per hundredweight if calves come in heavy; there's no corresponding slide for light calves. The slide takes some of the guesswork out of a contract and protects the buyer, which theoretically disposes them to contract for future delivery. Sam and Lusty contracted for $1.06 per pound at 535 pounds (the average weight of Sam's calves last year), so the slide reduced the eventual price per pound. Also, the actual weight of 653 pounds was reduced by a 3 percent "shrink," a normal practice to compensate for cattle that may have watered recently, increasing their weight on the scales. Small ranchers without livestock scales must haul their calves to auction barns, where no shrink is applied. Thus the calves' average net weight was 634 pounds, with a payout of $621 per head. As Jesús said when we drove the herd toward the pens this morning, "Mucho dinero." I agreed, "Mucho dinero, sí." More per head than Sam has ever received for his calves, only because he contracted back in the summer. After September 11 the market plummeted. These same calves on the market now might bring only 78 cents per pound.

After weighing the calves, we loaded them into the trucks while Sam and Lusty did their paperwork. Then they went to look at Sam's bred heifers, ones he held onto last year and hopes to sell as replacement heifers, for perhaps $875 a head. Lusty may have a buyer. The other calves we sorted—Herefords and small calves—aren't very promising, and Sam will sell them through an auction barn. The loads took off— *¡Adiós, niños!*

We turned our attention to the mamas, who stood in the catch pen, lowing for their departed babies. We moved the cows into the sorting pens and started them moving through the working chute, up to the portable squeeze chute. There Pete or Jesús caught the cow in the head gate for testing and vaccinations. Herman pulled on overalls and an arm-length plastic OB glove, which he doused with lubricating oil. Preg checking requires both feel and mettle: you insert your gloved hand into the cow's anus and follow the rectum back and downward, where you can then determine if a fetus is present in the adjoining womb. Not work for the squeamish. Herman would slip his arm in up to his shoulder, feel around a moment or two, and call out either "Bred" or "Open." Sometimes it's hard to tell, because a cow may be "lightly" bred. Each open cow is marked on the back with an orange grease pencil—"O" for open. Open and bred cows are released into different pens after testing, destined either to return to the pasture or to travel to the auction barn. A number of additional cows were culled before we started testing, many of them Herefords with bad eyes or bags, some simply old cows. Culls and open cows require ranchers to save or buy replacement heifer calves every year, selected to improve the herd as they replenish breeding stock. Sam keeps only black and black whiteface heifers as replacements; Hereford cows are slowly disappearing from his herd. To keep the fifty-fifty Angus-Hereford cross he prefers for his mama cows, however, he'll eventually have to buy Hereford replacement heifers, chosen for pigmentation that will protect their often vulnerable eyes and bags.

Between Bill Mitchell and Herman—who can check with either hand, a valuable skill with lots of cows to test, keeping his arm from getting tired—we worked our way patiently and methodically through nearly two hundred cows. You can't go any faster than the man pregnancy-testing. My job was to move cattle from sorting pens into the chutes, feeding a steady diet to the cowboys working up front. Most of Sam's cows work easily if you handle them slowly and don't crowd them much, though you always get an occasional rogue, a bitch who gets hot and won't hit the hole. Sam calls such critters "silly." A hotshot (an electric prod) does the trick then, but you've got to apply pressure when the cow is headed in the right direction or else you just juice her up even more. Bad cowpunchers, with little patience or a mean streak, will make cows silly and the work more difficult. These are Sam's mamas; I showed them respect.

We got to the last cow about noon, then rode back to headquarters for a big meal: brisket and broccoli and bread and salad and blueberry dessert.

The visitors all took off, we took a little siesta, then got ready for tomorrow, moving the squeeze chute to the pens up on the mesa, restocking our supplies, moving those twenty-four weaned steers to the corrals here at headquarters.

October 25, 2001—Thursday night. Other than a brief pause midday for coffee and a cookie, it was cattle work all day. We trailered our horses up to Three Corners pens while Herman took a feed truck to the east side of South Sabino to honk up any cows and calves there—they will always come for cake. We met him midway across the pasture and rode drag while Herman suckered the cattle toward the trap. A fifty-three-hundred-acre pasture is big for four men to gather horseback, and no one felt bad about the ploy, which worked quite well, thank you. I rode the middle, Jesús took the north side, and Sam checked the cedars along the mesa rim. We pushed a dozen pairs to Herman and the herd, then moved the whole bunch into the trap. From the gate, cattle can't see the pens at all—a nifty piece of work that makes driving them much easier. Once in the trap we sorted off Hereford cows with black whiteface calves; Sam wanted to keep only half-Hereford, half-Angus heifer calves for true black baldy replacements. Sam made the cut on Dunnie, with me on Pajamas, Herman on Streak, Jesús on Smoky. We sorted out sixty-four top pairs, including Herman's great big Angus cow and her black calf. He now has four Angus pairs and I don't know what else.

We drove the sixty-four pairs into the pens, stripped the calves, and worked them in the squeeze chute: ear-tagging, vaccinating, applying wormer. We finished by noon, took a coffee break to counter the cold wind that blew all morning, then worked the cows: pregnancy testing, vaccinating, worming. Out of sixty-four cows, fifty-nine were bred and kept—a healthy 92 percent conception rate. Despite other problems, Herefords tend to conceive readily. Three cows were open, and two with bad eyes weren't even tested—those five will head for the sale barn.

Tomorrow we'll have about 140 pairs—twice what we worked today. Lots of cows for Herman to preg-check, but an easier gather, no sorting, no ear tagging. We should come through fine. If nothing wrecks. Today we came close, when the cows knocked down a section of corral fence. Three Corners is built of wood and wire, and a couple of old posts snapped off at the ground when cattle crowded the outside fence. Though nothing got out, we had to rig a quick fix. We righted the fence, used the pickup to prop up the broken poles, and set an old gate in place to hold the fort.

Self-reliant, resourceful, gritty—that's a Cowden cowpuncher. Especially gritty, with that wind blowing dust down our craws.

After working the cows, we moved them across the trap into Conchas; with the trap between them, the cows and their weaned calves have two good fences keeping them apart. It will be noisy up there tonight, lots of mamas and babies bawling. A few may get back together, but not many, and not for long.

We came back to headquarters, unsaddled, had a bite to eat, then Sam and I resaddled the horses for the kids and hauled them into town for 4-H. It wasn't a good time for Sam to leave the ranch, but there's almost never a good time and Herman and Jesús could prepare for tomorrow—repair the corral, restock medicine and supplies, check on the cows and calves. Although inconvenient, the trip shows how much value the Cowdens place on family. Also, the woman who teaches 4-H riding owns the mare Sam's been considering for Hannah, and he needed to make a decision. First we took the mare to the vet, who comes to Santa Rosa from Las Vegas once a week. Sam had scheduled an appointment, but the vet was late and a backlog of sick and injured stock—mostly horses—waited at the rodeo grounds. As the kids rode in the arena, Sam visited with neighbors, and I rested in the pickup, pulling my hat over my eyes and dropping off briefly. Sam woke me with the vet's report: the mare is nearly blind in one eye and has a cataract developing in the other. No sale.

After 4-H ended, Sam gave the bad news and the mare back to Gretchen. We left Santa Rosa about 7:30 and were home by 8:15 to unsaddle the horses, unload groceries, and make our way toward bed. The weaned calves here in the corrals aren't making near as much racket as last night, but nothing—not calves, not a chorus of coyotes—nothing would keep this tuckered puncher awake.

October 26, 2001—We worked 136 calves and their mothers today, last day of my fall work. A good day, not too cold after the early chill, not too windy or dusty. The calves didn't get ear tags, so we raced through them; the cows went fine except for one or two locos. Tomorrow I drive to Santa Fe for a couple of days, then back East.

I pregnancy-tested the very last cow, and the one before, which Herman checked first; both were bred. Herman was going to double-check my final call, but I assured him and Sam the cow was bred, because she was. I know—I thrust my arm down inside her and felt the fetus. "Write down her number," I said, and they did, though no one will remember next spring, certainly not me. "If nothing else," I told Sam, "you can blame every open cow next year on your old primo." Sam and Herman thought that was a good plan. Each year some cows do not deliver calves, either because they weren't actually bred or because they abort sometime after pregnancy testing, usually due to disease. Because it's invasive, testing itself can provoke abortion. On a ranch this size, it's hard to determine why pregnancy-tested cows do not produce calves.

I had never preg-checked a cow before, so Herman gave me instructions. First I slipped on an arm-length plastic glove, secured it at my shoulder with a rubber band, then treated it with lubricating oil—cowboy KY Jelly. Then I came up behind the cow that was standing in the squeeze chute, raised her tail, and pointed my fingers and thumb forward to make a narrow probe that I slid past her anus. I made a fist inside her to straighten out the rectum, and extended my arm deep inside her until I reached what Herman called "a second asshole." I'm not sure what that opening was—the pelvis? After negotiating that, I slid my hand to the right and then down into the belly of the beast, at which point I was able to feel the fetus in the womb next to the cow's guts. "How big will it be?" I asked Herman. "Oh, about the size of a rabbit." Depending on how long the cow has been bred, you feel for different parts of the reproductive system and can even feel the ribs or hooves of a large fetus. And that's what I felt so clearly with the last two cows. It's an odd sensation, surely, to run your entire arm into the guts of a cow and feel for life inside. Once you overcome the indelicacy of the moment, it's not unpleasant—warm and tight—but I can see why Herman gets tired checking 150 cows. They hardly ever attempt to kick you, but sometimes they'll back up on you, jamming you against the chute. Can you blame them? In most cases, they stand reasonably peacefully while you make your examination. Then you withdraw your arm, plastic sleeve coated with feces, mark your cow bred or open, and move on—*Next!* When they weren't giving instructions, Sam and Herman razzed me with the fundamental, earthy humor of cattle country: "She likes you, primo! She likes you a lot!"

And I liked her and the little calf inside, five months away from this world of milk and grass, light and wind, plains and mountains, horses and cowboys. All winter I will think of them, destined for a birth in spring somewhere on this beautiful range.

We wrapped up things at the pens and drove back to headquarters with a sense of accomplishment and relief. I unsaddled Hollywood for the last time this year, breathing in the rich scent of sweat and leather and horseshit. We had a cold beer in the Bull Pen, sitting tired and grimy while Sam and

Herman checked their tally books, counting what was where on the ranch. The work's *not* all done this fall, just this week. There are still cattle to scatter to various pastures, newly weaned calves to tend, water and fences and equipment requiring attention, neighbors to help. But everyone gets the weekend off. I leave the ranch for Santa Fe, Kathy and the kids go to Lubbock, Sam goes to Clovis, and Jesús—though he doesn't yet know it—heads back to Mexico. No winter work here for Jesús, but I bet he's ready to go. Tomorrow is Herman's birthday and he gets to sleep in.

Now we are all in our respective places: Sam in the ranch house, Herman in his, Jesús at Alamito, Michael in the bunkhouse. That hot shower felt good; this whiskey tastes great. Soon I'll pack my "war bag," as old-time cowhands called their duffle, back when their worldly possessions, their "plunder," totaled little more than a change of clothes, some tack, a saddle. Among men, cowboys seemed less materialistic than most, in part because they valued what they did more than what they had, and in part because few goods were available on western ranges. Although cattlemen were different, with families and herds and holdings, in the West the democratic ideal was not entirely sentimental and often you could not distinguish ranchman from ranch hand—both were in the saddle at dawn, with much the same work before them. Sam is making up a grand tally for me, though, so I can see how things stand now, however different they may be tomorrow.

🌿 🌿 🌿

October 27, 2001—I drove to Santa Fe this morning, after saying adiós to Sam and Kathy, the kids, and Jesús. Herman was still sleeping off the rigors of the week and maybe the effects of the whiskey we drank last night. He came over to the bunkhouse with a bottle of Gentleman Jack, and we sipped a little and shot the bull. I've known Herman for nearly forty years, since Sam was in diapers, but his character is clearer now. As much as anyone, Herman embodies the Lazy 6.

Today is his birthday: forty-seven years old, born in 1954, west of Santa Rosa "in San Ignacio somewhere, I don't know." Second-to-last of eleven children—nine boys, two girls—he grew up poor and hardworking, married and took a job on the Cowden Ranch immediately after high school. He and Debbie have two sons, both raised on the ranch but neither one interested in cattle work. I asked him about that once and gathered that he wanted them to have more opportunities than he did. Yet Herman seems more at peace with his condition than almost everyone I know, though I'm sure he'd like to make more money and would prefer less work, or at least more help with routine jobs he's been doing for thirty years. Sam's well aware of that and does what he can to provide some sense of progress or relief, which Sam also would welcome. Ranch life requires the stoic persistence that Sam and Herman share; both are patient, good-humored, bright—the list of common virtues goes on. "We hardly have to talk sometimes," Sam says. "We usually know what the other one's going to do in any situation."

Last night at the kitchen table it was just me and Herman, with his thick glasses and Fu Manchu mustache. He drives around the ranch feeding cattle and listening to Paul Harvey on the radio, whistling while he works. Once he and I pulled the windmill at Antbed, Herman climbing the tower to secure the blades and to hang block and tackle we used to draw up the lengths of pipe from the casing. Up there framed against the New Mexico sky at sunset, Herman seemed perfectly in place and at peace. It's his habit to flash the peace sign at the Cowden kids—Abby, Hannah, and Guyito will forever remember him as a profound presence, an uncle, *un tío*—colloquial Spanish for an older man, uncle in spirit if not in fact. Herman echoes the Cowdens in generations past, who considered everyone close to the ranch an uncle. I'm sure Herman and Sam feel much the same about Leonard Lujan. *Abuelo,* "grandfather," is used in a similar way, though more to suggest ancestor; among Indians it can refer to the creator. This speaks to an enduring relationship arising out of the land itself and to labors associated with that land, whatever the culture—Anglo, Spanish, Indian. That's part of the great appeal and mystery of New Mexico.

Now I'm in Santa Fe—same state, different state of mind. Money, honey. Tourists swarm over the Plaza, eyeing the wares of Indians under the portal of the Governors' Palace or dropping into shops that trade on art and history made centuries ago in pueblos and haciendas, along the river, in the mountains, and on high lonesome plains like the Cowden Range. You can get your cowboy hat, Spanish weaving, Indian turquoise and silver, whatever costume suits you. This morning when I left the ranch, on the coatrack in the bunkhouse I hung a birthday present for Herman—a new Carhart jacket he admired when I first arrived. "Say Mike, that's a nice coat. Better not forget it in the pickup," he teased. "It may disappear." Last night we took one final shot—*¡Salud!*—and I let Herman know that I *would* forget it, and let him know exactly where.

Here's the tally Sam gave me, dated October 26, 2001, after the calves had shipped, pastures and numbers that now have grass and water tanks, flesh and blood, in my mind:

Conchas	186 cows	
Bogie	171	
House Trap	27	
3 Corners Trap	18	
South Sabino, heifer calves	201	
North Sabino, heifer yearlings	152	
Volts	45	
Creek pairs	125?	40 bulls
House, steer calves	25	11 horses
Mesa	40 corrientes	

That totals over a thousand head of cattle. On 50,000 acres, that's a stocking rate of one unit per 50 acres, typical for this country. People back East drop their jaws at 50,000 acres but don't realize how much land is required in semiarid ranges. In fact, western cattlemen almost always speak in sections—each one 640 acres—when they convey the size of ranches; acres are small, eastern measures. Though the Cowden Land and Cattle Company is one of the larger outfits in New Mexico, operations on a much smaller scale would be difficult and not so productive. When you see Sam Cowden and Herman Martinez operating the ranch without additional regular help, you appreciate *how* productive—their effort provides beef each year for approximately three thousand Americans, based on per capita consumption. This is a working ranch.

Yet an economic snapshot is not the big picture. Cattle culture today—as I see it alive on the Cowden family ranch—is my real interest. I've wanted to look at it from the inside out, with a historical context that places the Cowdens surely in their own world, whatever marginal position they occupy in current American life. This evening I went with friends to the Cowgirl Hall of Fame, a popular night spot adorned with nostalgic western trimmings—old ropes and hats and ranch paraphernalia—like the real things hanging in the Bull Pen on the Cowden Ranch. Nothing wrong with a honky-tonk, but after Sam Shepard dropped in for a while to hear the music, I realized that the joint was Hollywood. And I kept thinking how the important Hollywood is a sorrel quarter horse.

October 31, 2001—Happy Halloween from thirty-five thousand feet. We took off north from Albuquerque and turned east, between Sandia Peak and the Sangre de Cristos, Coronado's route four and a half centuries ago when he searched for the Seven Cities of Gold. By the time we passed over the ranch, cloud cover obscured the ground. Ahead are the Great Plains, Mississippi River valley, East Coast—I should be home in time for kids in costumes knocking on my door. If any come.

In Mexico—what my grandparents always called Old Mexico—they celebrate El Día de los Muertos, the Day of the Dead, November 1 and 2. All Saints' Day and All Souls' Day of the Catholic church have been transformed into a uniquely Mexican tradition: altars with food and drink, votive candles, and incense to greet the returning dead; candlelight vigils in cemeteries; fireworks and music and flowers. Toy and edible skulls, skeletons, and coffins mock death at the same time they acknowledge its presence. No one turns away from the dead; they embrace the souls of ancestors, children, friends. It's a lively and lovely celebration. Puritan New England, locked in its cold manners, has no equivalent. I'll miss New Mexico.

Sunday at the Tesuque Pueblo Flea Market, I got myself a cowhide rug, a nice big Hereford hide. Another Hereford hide was available, and I called Sam to see if he wanted it. They have a couple of hides from his own cattle, but it costs more to have a hide tanned than it does to buy one, and that tight Sam Cowden is always ready for a bargain. The hides are from Brazil, where most cattle resemble Longhorns, multicolored rather than uniform like British breeds. They split the hides and get one with hair and one suede hide; that's why they cost less. The dealer is shipping the hides to both Sam and me. To make sure we get the right ones, I wrote brands on the back: "Lazy 6" for Sam's and "JAL" on mine. I'm also thinking of actually burning the hide with one of the branding irons I have at home.

I'm over most of the effects of frequent, if not always hard, riding this fall. Only my left ankle aches—ankles, knees, and groin were the constant pressure points. I'm glad to have survived intact, although I wasn't worried that I'd get hurt. It can happen, though. On our final day one of the last cows, snuffy from the long wait and frequent handling, tried to climb out of the chute while Herman was preg-checking the cow ahead. She managed to rap Herman on the top of his head—hello!—not often a spot where you get kicked. Sam got kicked a number of times, and once a cow jammed my hand against the squeeze chute when I was giving a shot—my only real ding. Sun and wind burned and tanned my skin, which left me looking healthy. Lean, mean cowboy machine Michael.

Last leg. Which reminds me of the night Leonard Lujan stayed over and bunked on the bed next to mine. We'd all been sipping whiskey and spinning windies in the bunkhouse—he wasn't about to drive home. When we were ready to turn in, Leonard retrieved his crutch from his truck, laid it next to his bed, undressed, unstrapped his wooden leg and set that down too. Old one-legged Apache cowboy was ready for the night. It wasn't what I'd call restful—Leonard got up two or three times, using his crutch to negotiate the short trip to take a leak. That one night led me to appreciate all the more Leonard's outwardly modest yet truly remarkable life. Leonard, Herman, Sam, Rooster, Big Guy, Big Daddy, William Hamby Cowden, all those men and all the women with them—Selena, Debbie, Kathy, Jean, Mushy, Moner, Carrie Liddon Cowden—deserve my attention and appreciation. Failure to recognize the value of their lives is a form of blindness.

Which further reminds me. When we were gathering the cows and calves in Conchas last spring, we turned up a blind calf—Sam says they have one every year or two. It was a sad sight, the otherwise healthy calf stumbling along, bawling for his mother, following her scent and lowing as they moved across the pasture. As they did, the calf stumbled into one cactus after another, cholla spines gathering in his muzzle, his nose and lips pricked with blood. All the cowboys moved him along gently but occasionally he'd make a turn, lose his mother, and scramble away in the wrong direction, panicked until mama came to retrieve him. Once we had the herd in the branding pens and began working the calves, the blind calf was completely lost, frozen with fear when not careening around the pen. I lost sight of him as the morning's work went on, but he turned up again with a bang. After he was roped and dragged, flanked and worked and released, the blind calf ran straight into the metal table that held our medicine and supplies and water cooler. Ran straight through it, in fact, banging between the legs and scrambling out the other side as he overturned everything in his black path—he missed running into the red-hot branding pot by about two feet. People unaccustomed to the some-times harsh life of ranching would have been horrified; even the old hands had to shake their heads.

Equally sad as the calf's experience that day is the thought of living in beautiful country he will never see. I realize I'm projecting human values onto cattle—do animals really give a damn about scenery?—yet think of the loss, the aesthetic deprivation. A sense of vision—literal and figurative—is one more gift the Cowden Ranch provides, and it helps everyone there weather years of isolated labor. You ride to the edge of a mesa, on a ranch your grandfather put together, looking for stray cows and calves branded with a Lazy 6. Arrayed before you are high plains and mountains so clarified by light you seem to see past the horizon, through time, back through generations that brought you here. "If the doors of perception were cleansed," wrote William Blake, "every thing would appear to man as it is, infinite."

They just announced our approach to Hartford. My ride for the brand is over. I'm going back home, boys. The work's all done this fall.

Notes

Roundup

5 *Their cattle ran*   Cochran, "Story of the Early Days."
7 *The Indians stole*   Ibid.

Tumbleweeds: Branding 2000

16 *Each clustered*   Horgan, *Great River*, 268–69.
17 *The quirky shape*   See M. Rock, "Anton Chico and Its Patent."
18 *Hurrah Creek*   See Simpson, *Report of Exploration*, 11.
20 *A quick cowboy glossary*   See Culley, *Cattle, Horses, and Men.*
20 *Rustle up a copy*   Adams, *Western Words.* When a waddy (old cowboy)
   snakes a hoolihan (throws his rope) around a snuffy (wild or spirited
   horse), slaps his tree on (saddles up), grabs the apple (saddle horn),
   and hazes (gently drives) some corrientes (Mexican cattle) toward a
   democrat pasture (formed by stringing a fence across the narrow mouth
   of a canyon), you're seeing pretty punchy stuff (authentic cowboy work).
21 *Say some historians*   See, for example, Morris, *El Llano Estacado,* 26.
25 *Southern culture*   See Meinig, Imperial *Texas,* and Terry Jordan, *Trails
   to Texas.*
26 *Willa Cather wrote*   Cather, *Death Comes for the Archbishop,* 95.
26 *On average*   National Oceanographic & Atmospheric Association
   Technical Memorandum NWS SR-193, National Weather Service and
   National Severe Storms Laboratory, 1959-1994. Ten percent of those
   struck are killed; ninety percent suffer long-term effects.
27 *Darling Wife*   Letter courtesy of William Holt Jowell, Midkiff, TX.
28 *In 1884 or '85*   Harper, *Eighty Years of Recollections.* George Harper
   interviews and writes about Benjamin F. Harper.

29 *So concealed*   Pedro de Castañeda, in Hammond and Rey, *Narratives*, 261.

31 *While the army*   Ibid., 238. A span is the distance between a man's extended thumb and little finger—about nine inches.

33 *In burning the grass*   Gray, *Survey*, 18–19.

33 *The old grass*   Gregg, *Commerce of the Prairies*, 193.

33 *We had just passed*   Ibid., 306–307.

39 *He cares about his ranches*   Singleton told the *Albuquerque Journal* in 1987: "We hope to put any profits back into the improvement of the ranches. . . . What I'm interested in is improving the property as much as possible. I'm dedicated to the improvement of the land. I think everybody who owns property has the same feeling."

39 *Wild cattle range*   Haley, *Charles Goodnight*, 423.

45 *Earmarks have long been used*   See F. Ward, *The Cowboy at Work*, 63–75.

Gone to Texas

48 *He would be a rash prophet*   Turner, *Frontier in American History*, 38.

49 *We went naked*   Cabeza de Vaca, *Adventures*, 92.

50 *According to the muster roll*   Hammond and Rey, *Narratives*, 8.

50 *The army departed*   Castañeda, in ibid., 235. For other possible routes, cf. Bolton, *Coronado*, and Morris, *El Llano Estacado*. The previous winter had been severe; May would have seen dramatic snowmelt from the Sangre de Cristos. Flocks of woolly sheep, in particular, could not have crossed the swollen river.

50 *Leaving at his back*   Bolton, *Coronado*, 243.

51 *There were seen also*   Castañeda, in Hammond and Rey, *Narratives*, 236–37.

51 *Who could believe*   Ibid., 278–79.

51 *Perhaps some modern cowboy*   Bolton, *Coronado*, 252.

52 *Over these plains*   Castañeda, in Hammond and Rey, *Narratives*, 261–62.

52 *Kiowa and Missouri Indians*   Webb, *Great Plains*, 57.

52 *The horse did not introduce*   Ibid., 58.

53 *In part forced*   Wallace and Hoebel, *Comanches*, 10–11.

54 *Backwoodsmen—a bold*   Fehrenbach, *Lone Star*, 162.

54 *A set of North American*   Ibid., 163.

54 *Ninety percent*   Ibid., 279.

55 *Last night was rather*   Peskin, *Volunteers*, 25.

55 *Consisted of low hills*   J. Smith, *War with Mexico*, 1:205. Letter of Lieutenant George B. McClellan.

55 *A spot fit only*   Ibid., 1:206. Their first inland stop was Camp Belknap.
55 *Every breath of air*   Ibid., 1:211.
56 *They had come for glory*   Ibid., 1:207.
56 *Although I know*   Scott, *Memoirs*, 424.
56 *The soldiers had learned*   J. Smith, 2:63–64.
57 *Hays' regiment*   Fehrenbach, *Lone Star*, 490–91.
58 *Annexation*   Richardson, Anderson, and Wallace, *Texas: The Lone Star State*, 181–82.
58 *Immigration into Texas*   Fehrenbach, *Lone Star*, 287.
59 *The Texas Piney Woods*   Terry Jordan, *Trails to Texas*, 107.

Tumbleweeds: May 2000

60 *Lubbock arose*   See Hinshaw, *Lea*, 67–82, and Whitlock, *Cowboy Life*, 3–8.
61 *Everywhere civilization*   Hinshaw, *Lea*, 82.
62 *Here's one discovery*   Mosley, *"Little Texas,"* 2.
62 *Causey killed more buffalo*   Hinshaw, *Lea*, 69.
63 *We lived off game*   Rankin interview.
65 *In those days girls*   Watson interview.
65 *Pioneer members*   Price, *Open Range Ranching*, 44.
66 *It seemed to me*   Watson interview.
67 *The Cowdens at that time*   Ibid.
71 *From* The Handbook of Texas   A premier Texas resource, available in print and online versions.
75 *It avails not*   Whitman, "Crossing Brooklyn Ferry."

Cowboys and Indians

76 *Texas was a new country*   Dodge, *Plains*, 148.
76 *For anyone who loved*   Cochran, "Story of the Early Days."
76 *It seems to have been designed*   Marcy, *Thirty Years of Army Life*, 167.
77 *Nights are cool*   Graves, *Goodbye to a River*, 3.
77 *There were no farms then*   Bell interview.
78 *Traveling through Texas*   Olmsted, *Journey through Texas*. Olmsted was commissioned to write articles for the *New York Times;* the book was published in 1857. He then began his career as a landscape architect.
78 *The enormous strength*   Fehrenbach, *Lone Star*, 301.
78 *The Slaughters' choice*   Murrah, *C. C. Slaughter*, 8.
79 *Unquestionably*   McConnell, *West Texas Frontier*, 170.

79 *The men looked*   Clarke, *Palo Pinto Story,* 18–21. The round-trip freight run took six weeks.

81 *The early settlers*   Cochran, "Story of the Early Days."

83 *My people have never first*   Brown, *Bury My Heart,* 242.

83 *As in the case*   Wallace and Hoebel, *Comanches,* ix.

84 *The grandest of exploits*   Dodge, *Plains,* 262.

84 *Gradually the Anglo-American*   Wallace and Hoebel, *Comanches,* 292.

84 *A great many of the first settlers*   Cochran, "Story of the Early Days."

85 *The agonizing screams*   Wilbarger, *Indian Depredations in Texas,* 516.

86 *At the same time*   Murrah, *C. C. Slaughter,* 10–11.

86 *The people on the Texas frontier*   Cochran, "Story of the Early Days."

87 *They would build*   Bell interview.

87 *There are at least 160*   Clarke, *Palo Pinto Story,* 53.

87 *In 1863*   Cochran, "Story of the Early Days." Another "Big Drift" occurred in West Texas during the winter of 1884–85.

88 *Each year, up to the time*   Webb, *Great Plains,* 212.

88 *Neglected to brand*   *Handbook of Texas,* 2:162.

88 *The domestic cattle of Texas*   Dodge, *Plains,* 148.

88 *With their steel hoofs*   Dobie, *Longhorns,* 42.

89 *It was not considered*   Cochran, "Story of the Early Days."

89 *Commenced their mavericking days*   Ibid.

89 *There had been no stealing*   Haley, *Charles Goodnight,* 100–101.

89 *After the work was over*   Cochran, "Story of the Early Days."

90 *The horse was both*   Richardson, *Comanche Barrier,* 7.

90 *My father loaned*   Cochran, "Story of the Early Days."

90 *Every moonshiney night*   Bell interview.

90 *Where all are such magnificent thieves*   Dodge, *Plains,* 401.

91 *My father and I went*   Cochran, "Story of the Early Days."

91 *After the collapse*   Richardson, *Comanche Barrier,* 148.

91 *I went with two six-shooters*   Bell interview.

91 *Captain Frank Cowden*   Cochran, "Story of the Early Days."

92 *The frontier family*   Fehrenbach, *Lone Star,* 90.

93 *Sept. Thurs. 2*   Koen, "Social and Economic History," app. B, 164–79. Baker diary also printed in the *Palo Pinto County Star,* 1944–45, from the manuscript of Mrs. E. B. Ritchie, Mineral Wells, TX.

95 *To the emigrant*   Ormsby, *Butterfield Overland Mail,* 62–63.

96 *It is generally believed*   Cochran, "Story of the Early Days."

96 *We went a little south*   Bell interview.

96 *Travel by the Emigrant Trail*   Haley, *Charles Goodnight,* 128.

96 *The graveyard*   Ibid.

97 *In the spring of 1870*   Cochran, "Story of the Early Days."

97 *In 1870, me and Mr. Cowden*   Bell interview.

97 *Watch the procession*   Turner, *Frontier in American History,* 12.

97 *During the great slaughter*   Hornaday, *Extermination of the American Bison,* 501.

98 *These hunters have done more*   Sheridan, quoted in Hinshaw, *Lea,* 68.

98 *The slaughter went on*   Hornaday, *Extermination of the American Bison,* 501.

98 *Chesley Dobbs was found*   Wilbarger, *Indian Depredations in Texas,* 513.

Tumbleweeds: January 2001

109 *Wm. Cowden came very near*   James Baker, quoted in Koen, "Social and Economic History," 168, 172, 173.

110 *There were 479*   Foreman, *Marcy and the Gold Seekers,* 142–43.

114 *Perhaps, no part*   Marcy, *Thirty Years of Army Life,* 27.

126 *Now I've got time*   Cf. Dobie, *Voice of the Coyote.*

Give Me Land, Lots of Land

130 *[Palo Pinto] was*   Wolcott interview.

130 *These great steppes*   Gregg, *Commerce of the Prairies,* 192.

130 *Crazy Mexican brands*   Ford, *Texas Cattle Brands,* xix–xx.

132 *Wagons loaded with great stacks*   See Dodge, *Plains,* and Hornaday, *Extermination of the American Bison.*

134 *A thriving frontier settlement*   *Palo Pinto County Star,* April 13, 1878.

135 *Light as air*   *Handbook of Texas Online,* s.v. "Barbed Wire." John W. (Bet-a-million) Gates was the salesman for the San Antonio demonstration.

136 *Of his devoted wife*   Sterling, "Family History," courtesy of William Holt Jowell, Midkiff, TX.

136 *In 1875 the oldest son*   See Cox, *Historical and Biographical Record,* 442.

136 *The* Dallas Morning News   See Clarke, *Palo Pinto Story,* 168–69.

137 *Then, into this climate*   See ibid., 87.

138 *Gun that shoots today*   Hinshaw, *Lea,* 67.

139 *We came upon the high banks*   Marcy, *Report,* 68–70.

140 *On their return*   Cochran, "Story of the Early Days." 27.

140 *The brand's origin remains*   *Handbook of Texas Online,* s.v. "Cowden Ranch."

141 *Later mistranslated*   Bolton, *Coronado,* 243.

141 *There was nothing but grass*   Hammond and Rey, *Narratives,* 186.

141 *The original term*   *The Handbook of Texas,* 2:70.

141 *We reached some plains*   Hammond and Rey, *Narratives,* 186.

141 *The most notable of the great plateaux*   Gregg, *Commerce of the Prairies,* 181.

142 *The formation of the Llano Estacado*   U.S. Congress, Senate, *Pacific Railroad Survey,* 2:10.

142 *Its sublimity arises*   Pike, *Prose Sketches and Poems,* 9.

143 *The JAL Ranch was started*   Cochran, letter to Haley.

143 *The whole surface of the country*   Marcy, *Report,* 62.

143 *Then come the white sand hills*   Michler, in U.S. Congress, Senate, *Report,* 29.

145 *These singular-looking hills*   Gray, *Survey,* 20.

147 *We were at the ranch*   Bessye Cowden Ward interview.

147 *I have never seen better grass*   Shafter, quoted in Carlson, "William R. Shafter," 46.

148 *The whole country*   Shafter, quoted in Smith, "To Save a Dune," 19–20.

148 *A range of low hills*   Ibid., 20.

149 *There were plenty of Indian signs*   Cochran, "Story of the Early Days."

149 *W. H. Cowden had heard*   Cochran, "Brief History."

150 *As yet no ranchman*   Webb, *Great Plains,* 229.

152 *There never was any trouble*   Cochran, letter to Haley.

152 *James Hinkle recalls*   Hinkle, *Early Days,* 6.

153 *There was a large family*   Noble, "History of Midland County."

153 *JAL cowboys could also run*   Cochran, "Story of the Early Days."

154 *The San Simons*   Mosley, *"Little Texas,"* 18–19.

Tumbleweeds: May 2001

158 *44. WILLIAM HENRY COWDEN*   Ford, *Texas Cattle Brands,* 225–26.

162 *The first brand they registered*   Earmarks and, infrequently, dewlaps and wattles, are used as companions to brands in establishing ownership of cattle. In *The Cowboy at Work,* Fay Ward identifies well over a hundred different possibilities (67–75). Sam Cowden's heifer calves are earmarked "V-underbit the right."

164 Spencer Collins, interview by the author, May 2001.

165 *One day's tally*   Tally book, Bob Beverly Collection, Haley Library.

166 *Another revelation*   Note cards, Bob Beverly Collection, Haley Library.

174 *2,150 feet underground*   Waste Isolation Pilot Plant, http://www. wipp.ws.

Sin Agua, No Hay Nada

176 *Rain for us made history*   Gilbert, *We Fed Them Cactus,* 12.

176 *A desolate waste*   Marcy, *Report,* 42.

176 *Their methods and equipment*   Mosley, *"Little Texas,"* 5.

177 *Robert Rankin remembers*   Rankin interview.

177 *If there is such a thing*   Cochran, "Story of the Early Days."

177 *They proved to be*   Mosley, *"Little Texas,"* 10.

178 *While wind pumps water*   Webb, *Great Plains,* 21–22.

179 *George Cowden said*   Cochran, "Story of the Early Days."

180 *When the Indians saw*   Cabeza de Vaca, *Adventures,* 105.

180 *It was still subject to drought*   Data from National Climatic Data Center, National Drought Mitigation Center, National Oceanic and Atmospheric Administration, and Western Regional Climate Center.

180 *The two years' drouth*   Cochran, "Story of the Early Days."

181 *The plains west of here*   Dearen, *Crossing Rio Pecos,* 27.

181 *Some homesteaders*   Hubbs interview. Hubbs became a West Texas/New Mexico developer. According to Gary Blocker, "Barney and Mr. Knight developed a water company . . . and they sunk about six wells into the edge of the sand and got a lot of water. At that time there was no prohibition as to what you could do with the water. You owned the land, you owned the water. So they started developing the Dollarhide. No water on the Texas side of the Dollarhide. I don't know how your family records show that, but a lot of places they just flat couldn't find windmill water. It's kind of like they said, 'OK, we'll let New Mexico have the water, and we'll take the oil.'"

181 *Soon after the turn*   Mosley, *"Little Texas,"* 19–20.

182 *Such troubles didn't plague*   Beverly interview.

182 *A 1971 appraisal*   Cole, appraisal. The appraisal states: "This report has been prepared to determine the Fair Market Value, surface only, of this property used as a ranching unit as of September 25, 1969."

183 *Geologist William Blake reported*   U.S. Congress, Senate, *Pacific Railroad Survey,* 2:16.

183 *No one ever went back*   Benedict, *Tenderfoot Kid,* 10.

183 *Other adverse factors*   Cole, 13.

183 *Reports from the well*   *Galveston Daily News,* January 12, 1901.

184 *Clean water became difficult*   Spindletop–Gladys City Boomtown Museum, http://www.spindletop.org, 2000.

184 *Once the streams pass*   Webb, *Great Plains,* 12.

184 *In the last year*   Handbook of *Texas Online*, s.v. "Oil and Gas Industry."

185 *No. 1 E. P. Cowden*   Myres, *Permian Basin*, 186.

185 *An outstanding field opened*   Ibid., 186–87.

185 *The applicant says*   Edwards letters courtesy of Jean Cowden.

186 *Their situation echoed*   Haley, *Charles Goodnight*, 142.

187 *The Pecos river is here clear*   U.S. Congress, Senate, *Pacific Railroad Survey*, 6:5.

187 *Along the Canadian River*   Simpson, *Report of Exploration*, 11.

187 *Marcy's initial diary entry*   Foreman, *Marcy and the Gold Seekers*, 243–45.

188 *Simpson added*   Simpson, *Report of Exploration*, 17–18. Also cf. Morris, *El Llano Estacado*, 186: "Typically, a flock of about a thousand sheep was entrusted to the care of a lowly pastor. Two or more *pastores* were supervised by a *vaquero*. *Vaqueros* worked under a *caporal*, who in turn reported to the *mayordomo*. The flocks left the mountains in late January, drifted down the Canadian River, and then ventured onto the plains for lambing and summer grazing. Gradually stone corrals and other practical traces of the Hispano stock-raising frontier began to mark the cultural landscape of *La Ceja* and the Llano."

188 *Every year large parties*   Gregg, *Commerce of the Prairies*, 86.

189 *Parties of Comancheros*   Ibid., 219.

189 *Whipple's exploration*   U.S. Congress, Senate, *Pacific Railroad Survey*, 1:1.

189 *Whipple's survey along the 35th parallel*   Ibid., 3:38–41. Cowden headquarters is at almost exactly the same latitude as Anton Chico. Las Chupinas, where Whipple camped the night before Anton Chico, is at the same latitude as the saddle from the Esteros headwaters into the Alamito drainage.

191 *A map of the right-of-way*   Indexed map of New Mexico (Chicago: Rand McNally and Co., 1879).

191 *Between [the Pecos*   Koogler, "Stock Ranges," 207.

191 *The village of Anton Chico*   See Rock, "Anton Chico and Its Patent."

Tumbleweeds: Branding 2001

219 *To put the matter briefly*   Webb, *Great Plains*, 56.

219 *A "plains" saying*   Dodge, *Plains*, 417.

220 *J. Frank Dobie praises*   Dobie, *Mustangs*, 8–9.

220 *The name of these Springs*   Gray, *Survey*, 10.

220 *Dodge related*   Dodge, *Plains*, 329–30.

221 *The only property*   Marcy, *Thirty Years of Army Life*, 22.

221 *Poor horsemanship*   McGuane, *Some Horses*, 18.

221 *That echoes*   Catlin, *Letters and Notes*, 74–75.

223 *The literal meaning*   Cf. Dobie, *Mustangs*, 25: "Ranching on the open range necessitated mounting the serf, the cow-worker, or *vaca*-worker— the vaquero from whom the cowboy of North America took not only the occupation but the techniques and diction of his occupation. *Peon* means literally pedestrian, man-on-foot, the antipodes of *caballero*, man-on-horseback. The vaqueroized peon had a different status from the peon slaving in mines or as household menial. Sometimes he took horses and rode away from ranch or mission back to freedom."

224 *On their Thoroughbreds*   Cf. Rollins, *Cowboy*, 275: "A horse that could travel notably far, particularly when at high speed, was termed a 'long horse.'"

226 *A couple of passages*   Catlin, *Letters and Notes*, 64.

227 *You have always heard it said*   Cochran, "Story of the Early Days."

228 *Riding is second nature*   Dodge, *Plains*, 417.

Home on the Range

231 *Living at best*   McCoy, *Historic Sketches*, 405.

233 *Where families in town*   Horgan, *Great River*, 268–269.

233 *Wives came from the same*   Dary, *Cowboy Culture*, 272. Dary echoes Atherton, *Cattle Kings*, 83–84: "As a rule, their wives came from the same social strata in which they had been reared, another factor which contributed to the stability of marriages in the cattle kingdom."

234 *In that day and time*   Rankin interview.

235 *For the ranchers' wives*   Dary, *Cowboy Culture*, 262.

235 *There was no water supply*   Rankin interview.

235 *They had the Legal Tender saloon*   Beverly interview.

236 *With the cattle business*   Cleaveland, *No Life for a Lady*, 161.

237 *As ranches developed*   Dary, *Cowboy Culture*, 279.

237 *It was this deadly staying at home*   Cleaveland, *No Life for a Lady*, 156.

237 *Worst of all to me*   Dorothy Ross, *Stranger to the Desert*, 1958; quoted in Jordan, *Cowgirls*, xix.

237 *Newly married Gladys Cowden*   Watson interview.

238 *The cowboy of the west*   Beverly, *Hobo of the Rangeland*.

238 *The monotony of the prairies*   Mosley, "Little Texas," 14.

238 *Work, skill, endurance*   Hinshaw, *Lea*, 98.

239 *How many calves could you rope*   See Hugh Campbell interview: "Pink Paschul and Barnes Tillas were the best ropers I ever knew. Tillas ran the Con [Quien] Sabe ranch. I saw them in a roping contest. Tillas roped 332 calves and Pink 333 without missing a rope."

239 *Seeing to it*   Cleaveland, *No Life for a Lady,* 177.

241 *As transportation became less precarious*   Ibid., 271.

242 *The dances started at sundown*   Ibid., 169, 172.

242 *Back in 1853*   Sherburne, John. *Through Indian Country to California. John P. Sherburne's Diary of the Whipple Expedition, 1853-1854.* Edited by Mary McDougall Gordon. Stanford, CA: Stanford University Press, 1988. 103-104.

Tumbleweeds: Shipping 2001

247 *Most of our real work*   Cleaveland, *No Life for a Lady,* 129.

264 *It was the intention*   Cochran, "Story of the Early Days."

Bibliography

Adams, Andy. *Log of a Cowboy: A Narrative of the Old Trail Days.* Boston: Houghton, Mifflin and Co., 1903.

Adams, Ramon F. *The Old Time Cowhand.* New York: Macmillan, 1961.

———. *The Rampaging Herd: A Bibliography of Books and Pamphlets on Men and Events in the Cattle Industry.* Norman: University of Oklahoma Press, 1959.

———. *Western Words. A Dictionary of the Range, Cow Camp, and Trail.* Norman: University of Oklahoma Press, 1944.

Andrews County Heritage Committee. *Andrews County History, 1876–1978.* Andrews, TX, 1978.

Atherton, Lewis. *The Cattle Kings.* Bloomington: Indiana University Press, 1961.

Austin, Orval H. *Jal, New Mexico: Tough as an Old Boot.* Jal, NM: Jal Record, 1976.

Bauer, K. Jack. *The Mexican War 1846–1848.* Lincoln: University of Nebraska Press, 1992.

Beck, Warren. *New Mexico: A History of Four Centuries.* Norman: University of Oklahoma Press, 1962.

Beck, Warren, and Ynez Haase. *Historical Atlas of the American West.* Norman: University of Oklahoma Press, 1989.

———. *Historical Atlas of New Mexico.* Norman: University of Oklahoma Press, 1969.

Bell, Irbin. Interview by J. Evetts Haley, March 18, 1927, El Paso, TX. Transcript. Panhandle-Plains Historical Museum, Canyon, TX.

Benedict, Carl Peters. *A Tenderfoot Kid on Gyp Water.* Lincoln: University of Nebraska Press, 1986. First published 1943 by Texas Folklore Society, Austin.

Beverly, Bob. *Hobo of the Rangeland.* Lovington, NM, n.d.

———. Interview by J. Evetts Haley. Transcript. Bob Beverly Collection, Haley Memorial Library and History Center.

Bieber, Ralph P. "Southwestern Trails to California." *Mississippi Valley Historical Review* 12(3) (December 1925).

Blevins, Win. *Dictionary of the American West.* Seattle: Sasquatch Books, 2001.

Bolton, Herbert. *Coronado, Knight of Pueblos and Plains.* New York: Whittlesey House, 1949.

Brooks, Connie. *The Last Cowboys: Closing the Open Range in Southeastern New Mexico, 1890s–1920s.* Albuquerque: University of New Mexico Press, 1993.

Brown, Dee. *Bury My Heart at Wounded Knee.* New York: Holt, Rinehart, and Winston, 1970.

Cabeza de Vaca, Álvar Núñez. *Adventures in the Unknown Interior of America.* Trans. and ed. Cyclone Covey. Albuquerque: University of New Mexico Press, 1998.

Campbell, Hugh. Interview, September 30, 1938. Transcript. U.S. Work Projects Administration, Federal Writers' Project (American Memory Collection, Folklore Project, Life Histories, 1936–39). Washington, D.C. Manuscript Division, U.S. Library of Congress.

Carlson, Paul H. "William R. Shafter, Black Troops, and the Opening of the Llano Estacado, 1870–1875." *Panhandle-Plains Historical Review* 47 (1974): 1–18.

Cather, Willa. *Death Comes for the Archbishop.* New York: Vintage, 1971. First published 1927 by Knopf, New York.

Catlin, George. *Letters and Notes on the North American Indians.* Vol. 2 (of 2 vols. in 1), *1832–1839.* North Dighton, MA: JG Press, 1995.

Cattle Raisers Association of Texas. *History of the Cattlemen of Texas.* N.p., 1914.

Clark, W. P. *The Indian Sign Language.* Lincoln: University of Nebraska Press, 1982. First published 1885.

Clarke, Mary Whatley. *The Palo Pinto Story.* Fort Worth: printed by Manney Co., 1956.

Cleaveland, Agnes Morley. *No Life for a Lady.* Lincoln: University of Nebraska Press, 1977. First published 1941 by Houghton Mifflin, New York.

Cochran, W. C. "Brief History of the Early Days in Midland." Manuscript, June 10, 1929. Guy Cowden private collection. Later published in *Midland Reporter-Telegram,* October 13, 1935.

———. Letter to J. Evetts Haley, August 7, 1926. Panhandle-Plains Historical Museum, Canyon, TX.

———. "Story of the Early Days, Indian Troubles, and Cattle Business of Palo Pinto and Adjoining Counties." Unpublished memoirs. Guy Cowden private collection; copy courtesy of Jean Cowden.

Cole, C. A., Jr. Appraisal by Westbrook-Cole Mortgage Company, 1971. Courtesy of James Tom.

Cormier, Steve. "Times Were Not Easy: A History of New Mexico Ranching and Its Culture, 1900–1960." PhD diss., University of New Mexico, Albuquerque, 1998.

Cowden, John B. *Southern Cowdens.* West Nashville, TN, 1933.

Cox, James. *Historical and Biographical Record of the Cattle Industry and the Cattlemen of Texas and Adjacent Territories.* Vol. 2, *Biographical.* New York: Antiquarian Press, 1959. First published 1895.

Culley, John H. (Jack). *Cattle, Horses, and Men of the Western Range.* Tucson: University of Arizona Press, 1984. Facsimile of 1st ed. (Los Angeles: Ward Ritchie Press, 1940).

Dary, David. *Cowboy Culture.* New York: Knopf, 1981.

Dearen, Patrick. *Crossing Rio Pecos.* Fort Worth: Texas Christian University Press, 1996.

Dobie, J. Frank. *Guide to Life and Literature of the Southwest.* Rev. and enl. ed. Dallas: Southern Methodist University Press, 1981. First published 1952.

———. *The Longhorns.* Austin: University of Texas Press, 1997. First published 1941 by Little, Brown, Boston.

———. *The Mustangs.* New York: Bramhall House, 1934.

———. *The Voice of the Coyote.* Boston: Little, Brown, 1949.

Dodge, Richard Irving. *The Plains of the Great West and Their Inhabitants, Being a Description of the Plains, Game, Indians, &c. of the Great North American Desert.* New York: G. P. Putnam's Sons, 1877.

Erickson, John R. *The Modern Cowboy.* Lincoln: University of Nebraska Press, 1981.

Fehrenbach, T. R. *Lone Star: A History of Texas and the Texans.* New York: American Legacy Press, 1983.

Ford, Gus. *Texas Cattle Brands: A Catalog of the Texas Centennial Exposition Exhibit, 1936.* Dallas: Clyde C. Cockrell Company, 1936.

Foreman, Grant. *Marcy and the Gold Seekers: The Journal of Captain R. B. Marcy, with an Account of the Gold Rush over the Southern Route.* Norman: University of Oklahoma Press, 1939.

Frazier, Ian. *Great Plains.* New York: Penguin, 1989.

Freeman, James W., ed. *Prose and Poetry of the Live Stock Industry of the United States, with Outlines of the Origin and Ancient History of Our Live Stock Animals.* Denver and Kansas City: National Live Stock Historical Association, 1905.

Frost, Max. *New Mexico: Its Resources, Climate, Geography, Geology, History, Statistics, Present Condition, and Future Prospects.* Arranged, comp., and ed. Secretary of the Bureau of Immigration. Santa Fe, NM: 1894.

Gilbert, Fabiola Cabeza de Baca. *We Fed Them Cactus.* Albuquerque: University of New Mexico Press, 1954.

Graves, John. *Goodbye to a River.* New York: Knopf, 1961.

Gray, A. B. *Survey of a Route on the 32nd Parallel for the Texas Western Railroad, 1854: The A. B. Gray Report and Including the Reminiscences of Peter R. Brady Who Accompanied the Expedition.* Ed. and with introd. and notes by L. R. Bailey. Los Angeles: Westernlore Press, 1963. First published 1856 by Wrightson and Co. as *Railroad Record* print.

Gregg, Josiah. *The Commerce of the Prairies, or The Journal of a Santa Fe Trader, during Eight Expeditions across the Great Western Prairies, and a Residence of Nearly Nine Years in Northern Mexico; Illustrated with Maps and Engravings.* Lincoln: University of Nebraska Press, 1967. First published 1844 by H. G. Langley, New York.

Griffin, John Howard. *Land of the High Sky.* Midland, TX: First National Bank of Midland, 1959.

Haley, J. Evetts. *Charles Goodnight: Cowman and Plainsman.* Norman: University of Oklahoma Press, 1949.

———. *The XIT Ranch of Texas, and the Early Days of the Llano Estacado.* Norman: University of Oklahoma Press, 1967. First published 1929, Chicago.

Hammond, George, and Agapito Rey. *Narratives of the Coronado Expedition 1540–1542.* Albuquerque: University of New Mexico Press, 1940.

The Handbook of Texas. Vols. 1 and 2. Ed. Walter Prescott Webb. Austin: Texas State Historical Association, 1952.

The Handbook of Texas. Vol. 3, *A Supplement.* Ed. Eldon Stephen Branda. Austin: Texas State Historical Association, 1976.

The Handbook of Texas Online. http://www.tsha.utexas.edu/handbook/online/index.html.

Harper, George D. *Eighty Years of Recollections.* Works Projects Administration C.P. no. 665-66-3-299. Transcript. Panhandle-Plains Historical Museum, Canyon, TX.

Hinkle, James F. *Early Days of a Cowboy on the Pecos.* Roswell, NM, 1937.

Hinshaw, Gil. *Lea, New Mexico's Last Frontier.* Hobbs, NM: Hobbs Daily News-Sun, 1976.

Holden, W. C. *Rollie Burns, or An Account of the Ranching Industry on the South Plains.* College Station: Texas A&M University Press, 1986. First published in 1932.

Horgan, Paul. *Great River: The Rio Grande in North American History.* Austin: Texas Monthly Press, 1984. First published in 1954 by Holt, Rinehart, and Winston, New York.

Hornaday, W. T. *The Extermination of the American Bison.* Washington: Smithsonian Institution, 1889.

Hough, Emerson. *The Story of the Cowboy.* New York: Grosset and Dunlap, 1897.

Hubbs, Barney. Interview by Richie Cravens, July 8, 1977. Recording. Oral History Collection, Southwest Center/Special Collections Library, Texas Tech University, Lubbock.

Hunter, Angela. "A Place Apart: New Mexico's 'Little Texas' from 1880–1920." M.A. thesis, University of New Mexico, Albuquerque, December 1997.

Hunter, Marvin. *The Trail Drivers of Texas: Interesting Sketches of Early Cowboys and Their Experiences on the Range and on the Trail during the Days That Tried Men's Souls, True Narratives Related by Real Cowpunchers and Men Who*

Fathered the Cattle Industry in Texas. Austin: University of Texas Press, 1985. Orig. comp. and ed. by J. Marvin Hunter and published under direction of George W. Saunders by Jackson Printing Co., San Antonio, ca. 1920.

Jordan, Teresa. *Cowgirls: Women of the American West.* New York: Anchor Press, 1982.

Jordan, Terry. *North American Cattle-Ranching Frontiers: Origin, Diffusion, and Differentiation.* Albuquerque: University of New Mexico Press, 1993.

———. *Trails to Texas: Southern Roots of Western Cattle Ranching.* Lincoln: University of Nebraska Press, 1981.

Kendall, Geo. Wilkins. *Narrative of the Texan Santa Fé Expedition: Comprising a Description of a Tour through Texas and across the Great Southwestern Prairies the Camanche and Caygüa Hunting-grounds: With an Account of the Sufferings from Want of Food, Losses from Hostile Indians, and Final Capture of the Texans, and Their March, as Prisoners, to the City of Mexico.* With historical introd. by Milo Milton Quaife. Chicago: R. R. Donnelley and Sons, 1929. First published 1844 by Harper and Bros., New York.

Koen, J. C. "A Social and Economic History of Palo Pinto County." M.A. thesis, Hardin-Simmons University, Abilene, TX, 1949. Southwest Center/Special Collections Library, Texas Tech University, Lubbock.

Koogler, J. H. "Stock Ranges." In *Illustrated New Mexico,* by W. G. Ritch. 5th ed. Santa Fe: Bureau of Immigration, 1885.

Las Vegas Title Guaranty Company. "Abstract of Title to a Portion of the Preston Beck, Jr. Land Grant, San Miguel and Guadalupe Counties, New Mexico." Final certificate no. 12.657-A, State of New Mexico, County of San Miguel. Filed November 2, 1948, for records through April 25, 1945.

Lea County Genealogical Society. *Then and Now—Lea County Families and History.* 2 vols. Lovington, NM, 1979.

Loomis, Noel M., and Abraham P. Nasair. *Pedro Vial and the Road to Santa Fe.* Norman: University of Oklahoma Press, 1967.

Marcy, Randolph B., Colonel. *The Prairie Traveler: A Hand-book for Overland Expeditions; With Maps, Illustrations, and Itineraries of the Principal Routes between the Mississippi and the Pacific; Published by Authority of the War Department.* Bedford, MA: Applewood Books, 1993. First published 1859 by Harper and Bros., New York.

———. *Report of Captain R. B. Marcy [on Route from Fort Smith to Santa Fe].* In U.S. Congress, House, *Route from Fort Smith to Santa Fe.*

———. *Thirty Years of Army Life on the Border.* Introd. by Edward S. Wallace. Philadelphia: Lippincott, 1963. First published 1866 by Harper and Bros., New York.

Matthews, Sallie Reynolds. *Interwoven: A Pioneer Chronicle.* College Station: Texas A&M University Press, 1982.

McConnell, Joseph Carroll. *The West Texas Frontier, or A Descriptive History of Early Times in Western Texas Containing an Accurate Account of Much Hitherto*

Unpublished History, Presenting for the First Time in Historic Form a Detailed Description of Old Forts, Indian Fights and Depredations, Indian Reservations, French and Spanish Activities, and Many Other Interesting Things. 2 vols. Jacksboro, TX: Gazette, 1933.

McCoy, Joseph. *Historic Sketches of the Cattle Trade of the West and Southwest.* Ed. Ralph Bieber. Lincoln: University of Nebraska Press, 1985. First published 1874 by Ramsey, Millett, and Hudson, Kansas City, MO.

McGuane, Thomas. *Some Horses.* New York: Vintage, 2000.

Meinig, Donald W. *Imperial Texas: An Interpretive Essay in Cultural Geography.* Austin: University of Texas Press, 1975.

Morris, John Miller. *El Llano Estacado: Exploration and Imagination on the High Plains of Texas and New Mexico, 1536–1860.* Austin: Texas State Historical Association, 1997.

Mosley, May Price. *"Little Texas" Beginnings in Southeastern New Mexico.* Roswell, NM: Hall-Poorbaugh Press, 1973.

Murrah, David. *C. C. Slaughter: Rancher, Banker, Baptist.* Austin: University of Texas Press, 1981.

Myres, Samuel D. *The Permian Basin: Petroleum Empire of the Southwest.* Vol. 2, *Era of Advancement.* El Paso: Permian Press, 1977.

Newcomb, W. W. *The Indians of Texas.* Austin: University of Texas Press, 1961.

Noble, Susie G. "History of Midland County." Typescript. Midland Public Library, Midland, TX, n.d.

Noyes, Stanley. *Los Comanches: The Horse People, 1751–1845.* Albuquerque: University of New Mexico Press, 1993.

Olmsted, Frederick Law. *A Journey through Texas; or, A Saddle-Trip on the Southwestern Frontier.* Austin: University of Texas Press, 1978. First published 1857 by Dix, Edwards, New York.

Ormsby, Waterman L. *The Butterfield Overland Mail: Only Through Passenger on the First Westbound Stage.* Ed. Lyle H. Wright and Josephine M. Bynum. San Marino, CA: Huntington Library, 1942.

Palo Pinto County Historical Association. *History of Palo Pinto County (Word of Mouth Family History).* Dallas: Taylor Publishing Co., 1978.

Palo Pinto County Star. Souvenir ed., May 10, 1957.

Parker, W. B. *Notes Taken during the Expedition Commanded by Capt. R. B. Marcy, USA, through Unexplored Texas, in the Summer and Fall of 1854.* Austin: Texas State Historical Association, 1984.

Peskin, Allan, ed. *Volunteers: The Mexican War Journals of Private Richard Coulter and Sergeant Thomas Barclay, Company E, Second Pennsylvania Infantry.* Kent, Ohio: Kent State University Press, 1991.

Pike, Albert. *Prose Sketches and Poems, Written in the Western Country.* Ed. David J. Weber. Albuquerque: Calvin Horn, 1967. First published 1834.

Price, Eugene H. *Open Range Ranching on the South Plains in the 1890s.* Clarendon, TX: Clarendon Press, 1967.

Rankin, Robert. Interview by Richie Cravens, Abilene, TX, February 14, 1976. Recording. Oral History Collection, Southwest Center/Special Collections Library, Texas Tech University, Lubbock.

Reese, William S. *Six Score: The 120 Best Books on the Range Cattle Industry.* New Haven, CT: William Reese Co., 1989.

Reeves, C. C. (Tex), C. C. Reeves Jr., and Judy A. Reeves. *The Ogallala Aquifer of the Southern High Plains.* Lubbock, TX: Estacado Books, 1996.

Richardson, Rupert Norval. *The Comanche Barrier to South Plains Settlement.* Ed. Kenneth R. Jacobs. Austin: Eakin Press, 1996. First published 1933.

Richardson, Rupert Norval, Adrian Anderson, and Ernest Wallace. *Texas: The Lone Star State.* Englewood Cliffs, NJ: Prentice-Hall, 1981.

Rock, James L., and W. I. Smith. *Southern and Western Texas Guide for 1878.* St. Louis: A. H. Granger, 1878.

Rock, Michael J. "Anton Chico and Its Patent." In *Spanish and Mexican Land Grants in New Mexico and Colorado,* ed. John R. Van Ness and Christine M. Van Ness, 86–91. Manhattan, KS: Sunflower University Press, 1980.

Roe, Frank Gilbert. *The North American Buffalo.* 2nd ed. Toronto: University of Toronto Press, 1970.

Rollins, Philip Aston. *The Cowboy: His Characteristics, His Equipment, and His Part in the Development of the West.* New York: Charles Scribner's Sons, 1922.

Schlosser, Eric. *Fast Food Nation: The Dark Side of the All-American Meal.* New York: Houghton Mifflin, 2001.

Scott, Winfield. *Memoirs of Lieutenant-General Scott, LL.D.* 2 vols. New York: Sheldon and Co., 1864.

Sherburne, John P. *Through Indian Country to California: John P. Sherburne's Diary of the Whipple Expedition, 1853–1854.* Ed. Mary McDougall Gordon. Stanford, CA: Stanford University Press, 1988.

Simpson, James H. *Report of Exploration and Survey of Route from Fort Smith, Arkansas, to Santa Fe, New Mexico, Made in 1849, by First Lieutenant James H. Simpson, Corps of Topographical Engineers. House Executive Document 12.* In U.S. Congress, House, *Route from Fort Smith to Santa Fe.*

Siringo, Charles. *A Texas Cowboy, or Fifteen Years on the Hurricane Deck of a Spanish Pony, Taken from Real Life.* New York: William Sloane Associates, 1950. First published 1886 by Siringo and Dobson, Chicago.

Slatta, Richard. *The Cowboy Encyclopedia.* New York: W. W. Norton, 1994.

Smith, Calvin B. "To Save a Dune." *Greater Llano Estacado Southwest Heritage* 4(1) (Summer 1985): 19–21.

Smith, Justin. *The War with Mexico.* 2 vols. New York: Macmillan, 1919.

Smythe, Henry. *Historical Sketch of Parker County and Weatherford, Texas.* St Louis: Louis C. Lavat, 1877.

Sonnichsen, C. L. *Cowboys and Cattle Kings: Life on the Range Today.* Norman: University of Oklahoma Press, 1950.

Spindletop–Gladys City Boomtown Museum. http://www.spindletop.org.

Sterling, Cynthia Liddon. "Family History." Collected 1900–1915. Copied by Julianan Cowden, 1961. Courtesy of William Jowell.

Turner, Frederick Jackson. *The Frontier in American History.* Tucson: University of Arizona Press, 1986. First published 1920 by H. Holt and Co., New York.

U.S. Congress. House of Representatives. *Route from Fort Smith to Santa Fe. Letter from the Secretary of War transmitting reports of Lt. Simpson and Capt. R. B. Marcy.* 31st Cong., 1st sess., Exec. Doc. 45. Washington, 1850.

U.S. Congress. Senate. *Pacific Railroad Survey. Reports of Explorations and Surveys to Ascertain the Most Practicable and Economical Route for a Railroad from the Mississippi River to the Pacific Ocean. Made Under the Direction of the Secretary of War, in 1853–4, According to Acts of Congress of March 3, 1853, May 31, 1854, and August 5, 1854.* 11 vols. 33rd Cong., 2nd sess., Exec. Doc. 78. Washington: Beverley Tucker, Printer, 1855.

———. *Report of Lieutenant [Nathaniel] Michler.* 31st Cong., 1st sess., Exec. Doc. 64.

Von-Maszewski, W. M., and Matthew E. Von-Maszewski. *Index to the Trail Drivers of Texas.* Houston, TX: Tortuga Press, 1983.

Wallace, Paul, and E. Adamson Hoebel. *The Comanches: Lords of the South Plains.* Norman: University of Oklahoma Press, 1952.

Wallis, George A. *Cattle Kings of the Staked Plains.* Dallas: American Guild Press, 1957.

Ward, Albert E., John D. Schelberg, and Jerold G. Widdison, eds. *Archaeological Investigations at Los Esteros Reservoir, Northeastern New Mexico.* Albuquerque, NM: Center for Anthropological Studies, 1987.

Ward, Bessye Cowden. Interview by Wally Jackson, Odessa College, May, 14, 1964. Recording. Oral History Collection, Southwest Center/Special Collections Library, Texas Tech University, Lubbock.

Ward, Fay E. *The Cowboy at Work: All About His Job and How He Does It.* Norman: University of Oklahoma Press, 1987. First published 1958 by Hastings House, New York.

Waste Isolation Pilot Project. http://www.wipp.ws.

Watson, Lyda. Interview by Bill Collins, Midland, TX, April 25, 1960. Midland County Historical Society recording. Oral History Collection, Southwest Center/Special Collections Library, Texas Tech University, Lubbock.

Webb, Walter Prescott. *The Great Plains.* New York: Ginn and Co., 1931.

Whipple, A. W., Lt. *A Pathfinder in the Southwest: The Itinerary of Lieutenant A. W. Whipple during His Explorations for a Railway Route from Fort Smith to Los Angeles in the Years 1853 and 1854.* Ed. and annotated by Grant Foreman. Norman: University of Oklahoma Press, 1941.

Whitlock, V. H. (Ol' Waddy). *Cowboy Life on the Llano Estacado.* Norman: University of Oklahoma Press, 1970.

Wilbarger, J. W. *Indian Depredations in Texas.* Austin, TX: Eakin Press, 1985. Facsimile of 1st ed. (Austin, TX: Hutchings Printing House, 1889).

Wolcott, Henry. Interview by J. Evetts Haley, October 17, 1926, Midland, TX. Transcript. Panhandle-Plains Historical Museum, Canyon, TX.

Index

www.ingramcontent.com/pod-product-compliance
Lightning Source LLC
Chambersburg PA
CBHW032344280326
41935CB00008B/440